Problem-Solving Guide with Solutions
to accompany

COLLEGE PHYSICS

Volume I

Timothy A. French
DePaul University

W. H. FREEMAN AND COMPANY
New York

© 2014 by W. H. Freeman and Company
All rights reserved.
Printed in the United States of America

ISBN-13: 978-1-4641-0137-3
ISBN-10: 1-4641-0137-X

First printing

W. H. Freeman and Company
41 Madison Avenue
New York, NY 10010
RG21 6XS England

www.whfreeman.com/physics

Contents

Volume I

1. Introduction to Physics 1
2. Linear Motion 6
3. Motion in Two or Three Dimensions 19
4. Forces and Motion I: Newton's Laws 32
5. Forces and Motion II: Applications 46
6. Work and Energy 58
7. Momentum, Collisions, and the Center of Mass 69
8. Rotational Motion 80
9. Elastic Properties of Matter: Stress and Strain 93
10. Gravitation 101
11. Fluids 109
12. Oscillations 120
13. Waves 131
14. Thermodynamics I 150
15. Thermodynamics II 163

About the Author

DR. TIMOTHY FRENCH

Dr. Timothy French is currently a Visiting Assistant Professor in the Department of Chemistry at DePaul University in Chicago, IL. Dr. French holds a BS in chemistry from Rensselaer Polytechnic Institute and a MS and PhD in chemistry from Yale University. His undergraduate research looked at how to assess problem-solving ability in introductory physics students, while his doctoral research was on terahertz spectroscopy. Prior to arriving at DePaul, he was a Preceptor in Chemistry and Chemical Biology for four years at Harvard University where he taught an introductory physics sequence with biological and medical applications, experimental physical chemistry, and introductory physical chemistry, as well as the introductory physics class at the Harvard Summer School. He was awarded the Harold T. White Prize for Excellence in Teaching from the Harvard Physics Department in 2009.

A Note from the Author

The Problem-Solving Guide with Solutions is more than a book of answers to physics problems. It's designed to help you learn the way physicists approach and solve problems.

Many students, when approaching a physics problem, will half-read the question, pick out some of the variables used in the problem statement, and then start blindly hunting for an equation that shares the same letters, regardless of its utility.

Physicists, on the other hand, approach problems very differently. They'll fully read the question, underlining important information as they go. Next they will draw their own picture, even if one is provided, in order to quickly and easily organize and process the information provided. Then they will consider the underlying concept at play ("Is this a conservation of momentum question? Is Newton's second law best used here?") before determining which equation will be the most helpful. The equations are all manipulated algebraically before putting in numbers and arriving at a numerical answer.

Students usually move on to the next problem at this point without performing the crucial final step—checking the answer for reasonability. Physicists make sure the answer has the correct units, check some limiting cases ("What happens if the angle goes to 0 degrees? 90 degrees?"), and perform order of magnitude estimations all in order to gain confidence in their final answer.

Each solution in this book is set up to mirror this process with three distinct parts: **Set Up**, **Solve**, and **Reflect**. The **Set Up** portion contains all of the logic behind starting the problem, from a rehash of the important information in the problem statement to the conceptual underpinnings necessary in understanding the solution. The Solve portion contains the algebraic steps used to arrive at the numerical solution; note that this step comes *after* the Set Up step. Finally, the Reflect step provides a "sanity check" for the answer—"Is this what we expected? Does it make sense with respect to our observations of and interactions with everyday life? Does this answer seem reasonable?"

Very often instructors expect students to become expert problem solvers on their own without instruction. By explicitly showing the steps physicists use when solving problems, I hope you will be able to mirror and internalize these steps and make your foray into physics a more enjoyable and successful experience.

—Tim

Get Help with Premium Multimedia Resources

One of the benefits technology brings us in education is the ability to visualize concepts, gain problem-support, and test our skills outside of the traditional pen-and-paper classroom method. With that in mind, W. H. Freeman has developed a series of media assets geared at reinforcing conceptual understanding and building problem-solving skills.

P'Casts are videos that emulate the face-to-face experience of watching an instructor work a problem. Using a virtual whiteboard, the P'Casts' tutors demonstrate the steps involved in solving key worked examples, while explaining concepts along the way. The worked examples were chosen with the input of physics students and instructors across the country. P'Casts can be viewed online or downloaded to portable media devices.

Interactive Exercises are active learning, problem-solving activities. Each Interactive Exercise consists of a parent problem accompanied by a Socratic-dialog "help" sequence designed to encourage critical thinking as users do a guided conceptual analysis before attempting the mathematics. Immediate feedback for both correct and incorrect responses is provided through each problem-solving step.

Picture Its help bring static figures from the text to life. By manipulating variables within each animated figure students visualize a variety of physics concepts. Approximately 50 activities are available.

These Premium Multimedia Resources are available to you in a few places:
- If you're using the printed text, you can purchase the Premium Media Resources for a small fee via the Book Companion Website (BCS) at www.whfreeman.com/collegephysics1e.
- If you've purchased the W. H. Freeman media-enhanced eBook, these resources are embedded directly into the chapters of the text and available on the BCS.
- If you're using an online homework system, such as WebAssign, these resources are integrated into your individual assignments and are available on the BCS.

And while all three of those places correlate the Premium Media Resources by chapter, we've gone one step further in the Problem-Solving Guide with Solutions and correlated each media resource by problem. Look for each relevant item called out after its corresponding problem:

    Get Help: Picture It - Adding and Subtracting Vectors

To get started using these exciting resources, log on to www.whfreeman.com/collegephysics1e!

Chapter 1
Introduction to Physics

Conceptual Questions

1.3 Use of the metric prefixes makes any numerical calculation much easier to follow. Instead of an obscure conversion (12 in/ft, 1760 yards/mi, 5280 ft/mi), simple powers of 10 make the transformations (10 mm/cm, 1000 m/km, 10^{-6} m/μm).

1.5 No. The equation "3 meters = 70 meters" has consistent units but it is false. The same goes for "1 = 2," which consistently has no units.

1.7 The SI unit for length is the meter; the SI unit for time is the second. Therefore, the SI unit for acceleration is the meter/(second)², or m/s².

Multiple-Choice Questions

1.13 C (10^4).

$$1 \text{ m}^2 \times \frac{100 \text{ cm}}{1 \text{ m}} \times \frac{100 \text{ cm}}{1 \text{ m}} = \boxed{10^4 \text{ cm}^2}$$

Get Help: P'Cast 1.1 – The World's Fastest Bird

1.17 C (1810). Both 25.8 and 70.0 have three significant figures. When multiplying quantities, the quantity with the fewest significant figures dictates the number of significant figures in the final answer. Multiplying 25.8 by 70.0 gives 1806, which has four significant figures. Our final answer must have three significant figures, so we round 1806 to 1810.

Get Help: P'Cast 1.3 – Combining Volumes

Estimation/Numerical Analysis

1.21 We can split an average student's daily water use into four categories: showering, cooking/drinking/hand-washing, flushing the toilet, and doing laundry. A person uses about 100 L of water when showering, about 10 L for cooking/drinking/hand-washing, about 24 L when flushing the toilet, and about 40 L when doing two loads of laundry. This works out to about 150–200 L of water per day.

1.25 We can estimate the number of cells in the human body by determining the mass of a cell and comparing it to the mass of a human. An average human male has a mass of 80 kg. A person is mainly water, so we can approximate the density of a human body (and its cells) as 1000 kg/m³. We are told that the volume of a cell is the same as a sphere with a radius of 10^{-5} m, or approximately 4×10^{-15} m³; the mass of a single cell is $4 \times 10^{-15} \text{ m}^3 \times \dfrac{1000 \text{ kg}}{1 \text{ m}^3} = 4 \times 10^{-12}$ kg. The number of cells in the body is then

$$n_{\text{cell}} = \dfrac{m_{\text{body}}}{m_{\text{cell}}} = \dfrac{80 \text{ kg}}{4 \times 10^{-12} \text{ kg}} = \boxed{2 \times 10^{13}}.$$

Problems

1.29

SET UP

A list of powers of 10 is given. We can use Table 1-3 in the text to determine the correct metric prefix associated with each factor. Eventually, knowing some of the more common prefixes will become second nature.

SOLVE

    A) kilo (k)   E) milli (m)

    B) giga (G)   F) pico (p)

    C) mega (M)   G) micro (μ)

    D) tera (T)   H) nano (n)

REFLECT

Whether or not the prefix is capitalized is important. *Mega-* and *milli-* both use the letter "m," but *mega-* is "M" and *milli-* is "m." Confusing these two will introduce an error of 10^9!

1.35

SET UP

This problem provides practice with metric unit conversions. We need to look up (and/or memorize) various conversions between SI units and hectares and liters. One hectare is equal to 10^4 m^2, and 1000 L is equal to 1 m^3. Another useful conversion is that 1 mL equals 1 cm^3.

SOLVE

A) $328 \text{ cm}^3 \times \dfrac{1 \text{ mL}}{1 \text{ cm}^3} \times \dfrac{1 \text{ L}}{1000 \text{ mL}} = \boxed{0.328 \text{ L}}$

B) $112 \text{ L} \times \dfrac{1 \text{ m}^3}{1000 \text{ L}} = \boxed{0.112 \text{ m}^3}$

C) $220 \text{ hectares} \times \dfrac{10^4 \text{ m}^2}{1 \text{ hectare}} = \boxed{2.2 \times 10^6 \text{ m}^2}$

D) $44300 \text{ m}^2 \times \dfrac{1 \text{ hectare}}{10^4 \text{ m}^2} = \boxed{4.43 \text{ hectares}}$

E) $225 \text{ L} \times \dfrac{1 \text{ m}^3}{1000 \text{ L}} = \boxed{0.225 \text{ m}^3}$

F) $17.2 \text{ hectare} \cdot \text{m} \times \dfrac{10^4 \text{ m}^2}{1 \text{ hectare}} \times \dfrac{10^3 \text{ L}}{1 \text{ m}^3} = \boxed{1.72 \times 10^8 \text{ L}}$

G) $2.253 \times 10^5 \text{ L} \times \dfrac{1 \text{ m}^3}{10^3 \text{ L}} \times \dfrac{1 \text{ hectare}}{10^4 \text{ m}^2} = \boxed{2.253 \times 10^{-2} \text{ hectare} \cdot \text{m}}$

H) $2000 \text{ m}^3 \times \dfrac{1000 \text{ L}}{1 \text{ m}^3} \times \dfrac{1000 \text{ mL}}{1 \text{ L}} = \boxed{2 \times 10^9 \text{ mL}}$

REFLECT

Rewriting the measurements and conversions in scientific notation makes the calculations simpler and helps give some physical intuition.

Get Help: P'Cast 1.1 – The World's Fastest Bird

1.39

SET UP

In the United States, fuel efficiency is reported in miles per gallon (or mpg). We are given fuel efficiency in kilometers per kilogram of fuel. We can use the conversions listed in the problem to convert this into mpg.

SOLVE

$$\dfrac{7.6 \text{ km}}{\text{kg}} \times \dfrac{0.729 \text{ kg}}{1 \text{ L}} \times \dfrac{3.785 \text{ L}}{1 \text{ gal}} \times \dfrac{1 \text{ mi}}{1.609 \text{ km}} = \boxed{13 \dfrac{\text{mi}}{\text{gal}}} \text{ or } \boxed{13 \text{ mpg}}$$

REFLECT

This would be the gas mileage for a cargo van or large SUV. A hybrid sedan would have a gas mileage of around 40 mpg.

Get Help: P'Cast 1.1 – The World's Fastest Bird

1.43

SET UP

We are asked to calculate sums and differences with the correct number of significant figures. When adding or subtracting quantities, remember that the quantity with the fewest *decimal places* (<u>not</u> significant figures) dictates the number of decimal places in the final answer. If necessary, we will need to round our answer to the correct number of decimal places.

SOLVE

A) $4.55 + 21.6 = \boxed{26.2}$ C) $71.1 + 3.70 = \boxed{74.8}$

B) $80.00 - 112.3 = \boxed{-32.3}$ D) $200 + 33.7 = \boxed{200}$

REFLECT

The answer to part D may seem weird, but 200 only has one significant figure—the "2." The hundreds place is, therefore, the smallest decimal place we are allowed.

Get Help: P'Cast 1.3 – Combining Volumes

1.45

SET UP

We are given an equation that describes the motion of an object and asked if it is dimensionally consistent; we need to make sure the dimensions on the left side equal the dimensions on the right side. The equation contains terms related to position, speed, and time. Position has dimensions of length; speed has dimensions of length per time; and time has dimensions of, well, time.

SOLVE

$$x = vt + x_0$$

$$[L] \stackrel{?}{=} \frac{[L]}{[T]}[T] + [L]$$

$$[L] = [L] + [L]$$

REFLECT

Dimensions are general (for example, length), while units are specific (for example, meters, inches, miles, furlongs).

1.49

SET UP

We need to check that the dimensions of the expected quantity match the dimensions of the units reported. Remember that dimensions should be given in terms of the fundamental quantity (see Table 1-1 in the text). For example, volume has dimensions of (length)^3.

SOLVE

A) Volume flow rate has dimensions of volume per time, or $\frac{[L]^3}{[T]}$, so $\frac{m^3}{s}$ is correct.

B) Height has dimensions of [L]. Units of m^2 are not correct.

C) A fortnight has dimensions of [T], while m/s are the units of speed. This statement is not correct.

D) Speed has dimensions of $\frac{[L]}{[T]}$, but $\frac{m}{s^2}$ are units of acceleration. This statement is not correct.

E) Weight is a force and has dimensions of $\frac{[M][L]}{[T]^2}$, and lb is an appropriate unit for force. This is correct.

F) Density has dimensions of mass per volume, or $\frac{[M]}{[L]^3}$, so $\frac{kg}{m^2}$ is not correct.

REFLECT

Only statements A and E are correct. Making sure your answer has the correction dimensions and units is an important last step in solving every problem.

1.55

SET UP

It takes about 1 min for all of the blood in a person's body to circulate through the heart. When the heart beats at a rate of 75 beats per minute, it pumps about 70 mL of blood per beat. Blood has a density of 1060 kg/m^3. We can use each of these relationships as a unit conversion to determine the total volume of blood in the body and the mass of blood pumped per beat. For the volume calculation, we will start with the fact that all of the blood takes 1 min to circulate through the body. Because we are interested in the mass per beat, we can start with 1 beat in part b.

SOLVE

Part a)

$$1 \text{ min} \times \frac{75 \text{ beats}}{1 \text{ min}} \times \frac{70 \text{ mL}}{1 \text{ beat}} \times \frac{1 \text{ L}}{1000 \text{ mL}} = \boxed{5 \text{ L}} \times \frac{1 \text{ m}^3}{1000 \text{ L}} = \boxed{5 \times 10^{-3} \text{ m}^3}$$

Part b)

$$1 \text{ beat} \times \frac{70 \text{ mL}}{1 \text{ beat}} \times \frac{1 \text{ L}}{1000 \text{ mL}} \times \frac{1 \text{ m}^3}{1000 \text{ L}} \times \frac{1060 \text{ kg}}{1 \text{ m}^3} = \boxed{0.07 \text{ kg}} \times \frac{1000 \text{ g}}{1 \text{ kg}} = \boxed{70 \text{ g}}$$

REFLECT

A volume of 5 L is about 10 pints, which is about average. For comparison, they take 1 pint when you donate blood.

Get Help: P'Cast 1.1 – The World's Fastest Bird

Chapter 2
Linear Motion

Conceptual Questions

2.1 An object will slow down when its acceleration vector points in the opposite direction to its velocity vector. Recall that acceleration is the change in velocity over the change in time.

2.5 Speed and velocity have the same SI units (m/s). Speed is the magnitude of the velocity vector. If the velocity points completely in the positive direction, then the two can be interchanged. If the velocity is not fixed in direction, then the correctly signed component of the velocity will need to be used to avoid confusion.

2.9 The average velocity of a moving object will be the same as the instantaneous velocity if the object is moving at a constant velocity (both magnitude and direction). Also, if the average velocity is taken over a period of constant acceleration, the instantaneous velocity will match it for one moment in the middle of that period.

2.13 Assuming the initial speed of the ball is the same in both cases, the velocity of the second ball will be the same as the velocity of the first ball. The upward trajectory of the first ball involves the ball going up and reversing itself back toward its initial location. At that point, the trajectories of the two balls are identical as the balls hit the ground.

Multiple-Choice Questions

2.16 D.

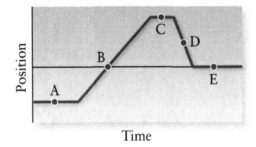

Figure 2-1 Problem 16

The slope of a position versus time plot gives information regarding the speed of the object. Point D has the largest slope, which means the object is moving the fastest there.

Get Help: P'Cast 2.3 – Decoding an x-t Graph

Estimation/Numerical Analysis

2.25 The fastest time for the 100-m race for men was set by Usain Bolt (9.58 s) and for women by Florence Griffith-Joyner (10.49 s). These times correspond to top speeds of 10.4 m/s and 9.533 m/s, respectively. A runner who falls in the mud will take much longer to come to rest than one who falls on a running track. Let's assume it takes 2.0 s to come to rest in the mud but only 0.50 s on a track. The magnitudes of the acceleration for each case are

$$a_{\text{Bolt,mud}} = \frac{\Delta v_x}{\Delta t} = \frac{10.4 \frac{m}{s}}{2.0 \text{ s}} = \boxed{5.2 \frac{m}{s^2}}$$

$$a_{\text{FloJo,mud}} = \frac{\Delta v_x}{\Delta t} = \frac{9.533 \frac{m}{s}}{2.0 \text{ s}} = \boxed{4.8 \frac{m}{s^2}}$$

$$a_{\text{Bolt,track}} = \frac{\Delta v_x}{\Delta t} = \frac{10.4 \frac{m}{s}}{0.50 \text{ s}} = \boxed{21 \frac{m}{s^2}}$$

$$a_{\text{FloJo,track}} = \frac{\Delta v_x}{\Delta t} = \frac{9.533 \frac{m}{s}}{0.50 \text{ s}} = \boxed{19 \frac{m}{s^2}}$$

Get Help: P'Cast 2.1 – Average Velocity

2.29 On the open sea, a cruise ship travels at a speed of approximately 10 m/s. We need to estimate the time it takes the ship to reach its cruising speed from rest. We should expect that it takes more than a few minutes but less than a full hour; let's estimate the time to be 0.5 h. The magnitude of the average acceleration of the cruise ship is the change in the ship's speed divided by the time interval over which that change occurs:

$$a_{\text{average},x} = \frac{\Delta v_x}{\Delta t} = \frac{10 \frac{m}{s}}{0.5 \text{ h}} \times \frac{1 \text{ h}}{3600 \text{ s}} = \boxed{0.006 \frac{m}{s^2}}$$

Get Help: P'Cast 2.7 – Motion with Constant Acceleration I: Cleared for Takeoff!
P'Cast 2.8 – Motion with Constant Acceleration II: Which Solution is Correct?

2.33 We can make a plot of position versus time and determine the equations for each region by manually calculating the slope and y intercepts or by using a computer program to fit the data in each time interval.

t(s)	x(m)
0	−12
1	−6
2	0
3	6
4	12
5	15
6	15
7	15
8	15
9	18
10	24
11	33
12	45
13	60
14	65
15	70
16	75
17	80
18	85
19	90
20	95
21	100
22	90
23	80
24	70
25	70

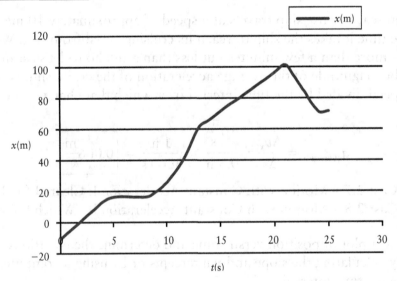

Figure 2-2 Problem 33

Between $t = 0$ s and $t = 4.5$ s:

$$x(t) = \frac{12 \text{ m} - (-12 \text{ m})}{4 \text{ s}}(t - (2 \text{ s})) = \left(6\frac{\text{m}}{\text{s}}\right)(t - (2 \text{ s}))$$

Between $t = 4.5$ s and $t = 8$ s: $x(t) = 15$ m.

Between $t = 8$ s and $t = 13$ s:

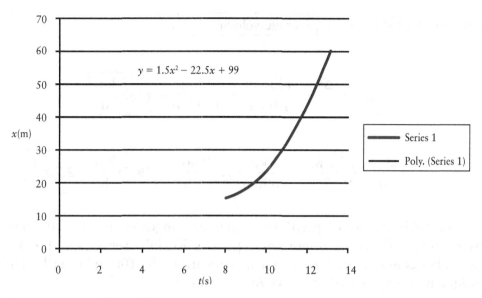

Figure 2-3 Problem 33

We can fit the data to a parabola in order to get position versus time in this region:

$$x(t) = \left(1.5\frac{\text{m}}{\text{s}^2}\right)t^2 - \left(22.5\frac{\text{m}}{\text{s}}\right)t + (99\text{ m}) = \left(1.5\frac{\text{m}}{\text{s}^2}\right)(t - 7.5\text{ s})^2 + (14.6\text{ m})$$

Between $t = 13$ s and $t = 21$ s:

$$x(t) = \frac{100\text{ m} - 60\text{ m}}{8\text{ s}}(t - 13\text{ s}) + (60\text{ m}) = \left(5\frac{\text{m}}{\text{s}}\right)(t - 13\text{ s}) + (60\text{ m})$$

Between $t = 21$ s and $t = 24$ s:

$$x(t) = \frac{70\text{ m} - 100\text{ m}}{3\text{ s}}(t - 21\text{ s}) + (100\text{ m}) = \left(-10\frac{\text{m}}{\text{s}}\right)(t - 21\text{ s}) + (100\text{ m})$$

Between $t = 24$ s and $t = 25$ s: $x(t) = 70$ m.

Get Help: P'Cast 2.7 – Motion with Constant Acceleration I: Cleared for Takeoff!
P'Cast 2.8 – Motion with Constant Acceleration II: Which Solution is Correct?

Problems

2.39

SET UP

Kevin swims 4000 m in 1.00 h. Because he ends at the same location as he starts, his total displacement is 0 m. This means his average velocity, which is his displacement divided by the time interval, is also zero. His average *speed*, on the other hand, is nonzero. The average speed takes the total distance covered into account.

SOLVE

Part a) Displacement = 0 m, so his $\boxed{\text{average velocity} = 0}$.

Part b)

$$v_{\text{average},x} = \frac{\Delta x}{\Delta t} = \frac{4000 \text{ m}}{1.00 \text{ h}} \times \frac{1 \text{ km}}{1000 \text{ m}} = \boxed{4.00 \frac{\text{km}}{\text{h}}}$$

Part c)

$$v_{\text{average},x} = \frac{\Delta x}{\Delta t} = \frac{25.0 \text{ m}}{9.27 \text{ s}} \times \frac{1 \text{ km}}{1000 \text{ m}} \times \frac{3600 \text{ s}}{1 \text{ h}} = \boxed{9.71 \frac{\text{km}}{\text{h}} = 2.70 \frac{\text{m}}{\text{s}}}$$

REFLECT

Usually the terms "velocity" and "speed" are used interchangeably in everyday English, but the distinction between the two is important in physics. Kevin's average speed of 4 km/h is about 2.5 mph. For comparison, Michael Phelps's record in the 100-m butterfly is 49.82 s, which gives him an average speed of 7.2 km/h or 4.5 mph.

Get Help: Interactive Exercise – Car Meets Train
P'Cast 2.1 – Average Velocity

2.43

SET UP

It takes a sober driver 0.320 s to hit the brakes, while a drunk driver takes 1.00 s to hit the brakes. In both cases the car is initially traveling at 90.0 km/h. Assuming it takes the same distance to come to a stop once the brakes are applied, the drunk driver travels for an extra (1.00 − 0.320) s = 0.680 s at 90.0 km/h before hitting the brakes. We can use the definition of average speed to calculate the extra distance the drunk driver travels.

SOLVE

$$v_{\text{average},x} = \frac{\Delta x}{\Delta t}$$

$$\Delta x = (v_{\text{average},x})(\Delta t) = \left(90.0 \frac{\text{km}}{\text{h}} \times \frac{1 \text{ h}}{3600 \text{ s}} \times \frac{1000 \text{ m}}{1 \text{ km}}\right)(0.680 \text{ s}) = \boxed{17.0 \text{ m}}$$

REFLECT

The impaired driver travels an extra distance of over 55 ft before applying the brakes.

2.47

SET UP

A runner starts from rest and reaches a top speed of 8.97 m/s. Her acceleration is 9.77 m/s², which is a constant. We know her initial speed, her final speed, and her acceleration, and we are interested in the time it takes for her to reach that speed. We can rearrange $v_x = v_{0x} + a_x t$ and solve for t.

SOLVE

$$v_x = v_{0x} + a_x t$$

$$t = \frac{v_x - v_{0x}}{a_x}$$

$$t = \frac{8.97\frac{m}{s} - 0}{9.77\frac{m}{s^2}} = \boxed{0.918 \text{ s}}$$

REFLECT

An acceleration of 9.77 m/s² means that her speed changes by 9.77 m/s every second. In this problem her speed changed a little less than that, so we expect the time elapsed to be a little less than a second.

2.51

SET UP

A Bugatti Veyron and Saleen S7 can accelerate from 0 to 60.0 mph in 2.40 s and 2.80 s, respectively. Assuming the acceleration of each car is constant, not only over that time interval but also for *any* time interval, we can approximate the acceleration of each car as the average acceleration. We first need to convert 60.0 mph into m/s in order to keep the units consistent. The acceleration is equal to the change in velocity divided by the change in time. We are interested in the distance each car travels when it accelerates from rest to 90.0 km/h. We know the initial speed, the final speed, and the acceleration, so we can use $v_x^2 = v_{0x}^2 + 2a_x(x - x_0)$ to calculate Δx for each car.

SOLVE

Converting to m/s:

$$60.0\frac{mi}{h} \times \frac{1.61 \text{ km}}{1 \text{ mi}} \times \frac{1 \text{ h}}{3600 \text{ s}} \times \frac{1000 \text{ m}}{1 \text{ km}} = 26.8\frac{m}{s}$$

Finding the acceleration:

$$a_{average,x} = \frac{\Delta v_x}{\Delta t}$$

$$a_{Bugatti,x} = \frac{26.8\frac{m}{s} - 0}{2.40 \text{ s}} = 11.2\frac{m}{s^2}$$

$$a_{Saleen,x} = \frac{26.8\frac{m}{s} - 0}{2.80 \text{ s}} = 9.57\frac{m}{s^2}$$

Finding the distance:

$$v_x^2 = v_{0x}^2 + 2a_x(x - x_0)$$

$$\Delta x = \frac{v_x^2 - v_{0x}^2}{2a_x}$$

$$\Delta x_{\text{Bugatti}} = \frac{\left(90.0\frac{\text{km}}{\text{h}} \times \frac{1000 \text{ m}}{1 \text{ km}} \times \frac{1 \text{ h}}{3600 \text{ s}}\right)^2 - (0)^2}{2\left(11.2\frac{\text{m}}{\text{s}^2}\right)} = \boxed{27.9 \text{ m}}$$

$$\Delta x_{\text{Saleen}} = \frac{\left(90.0\frac{\text{km}}{\text{h}} \times \frac{1000 \text{ m}}{1 \text{ km}} \times \frac{1 \text{ h}}{3600 \text{ s}}\right)^2 - (0)^2}{2\left(9.57\frac{\text{m}}{\text{s}^2}\right)} = \boxed{32.6 \text{ m}}$$

REFLECT

These values are around 92 ft and 108 ft, respectively. A speed limit of 90 km/h is around 55 mph. Most people don't drive expensive sports cars on the highway every day, so these distances are (obviously) much smaller than we should expect for most everyday cars.

Get Help: P'Cast 2.7 – Motion with Constant Acceleration I: Cleared for Takeoff!
P'Cast 2.8 – Motion with Constant Acceleration II: Which Solution is Correct?

2.53

SET UP

A sperm whale has an initial speed of 1.00 m/s and accelerates up to a final speed of 2.25 m/s at a constant rate of 0.100 m/s². Because we know the initial speed, the final speed, and the acceleration, we can calculate the distance over which the whale travels by rearranging $v_x^2 = v_{0x}^2 + 2a_x(x - x_0)$.

SOLVE

$$v_x^2 = v_{0x}^2 + 2a_x(x - x_0)$$

$$\Delta x = \frac{v_x^2 - v_{0x}^2}{2a_x} = \frac{\left(2.25\frac{\text{m}}{\text{s}}\right)^2 - \left(1.00\frac{\text{m}}{\text{s}}\right)^2}{2\left(0.100\frac{\text{m}}{\text{s}^2}\right)} = \boxed{20.3 \text{ m}}$$

REFLECT

This is a little longer than the average size of an adult male sperm whale (about 16 m). It will take the whale 12.5 s to speed up from 1 m/s to 2.25 m/s.

Get Help: P'Cast 2.7 – Motion with Constant Acceleration I: Cleared for Takeoff!
P'Cast 2.8 – Motion with Constant Acceleration II: Which Solution is Correct?

2.57

SET UP

Alex throws a ball straight down (toward $-y$) with an initial speed of $v_{A0y} = 4.00$ m/s from the top of a 50.0-m-tall tree. At the same instant, Gary throws a ball straight up (toward $+y$) with an initial speed of v_{G0y} at a height of 1.50 m off the ground. (We will consider the ground to be $y = 0$, which makes the initial position of Alex's ball $y_{A0} = 50.0$ m and the initial position of Gary's ball $y_{G0} = 1.50$ m). Once Alex and Gary throw their respective balls, their accelerations will be equal to $a_y = -g$. We want to know how fast Gary has to throw his ball such that the two balls cross paths at $y = 25.0$ m at the same time.

We are given more information about Alex's throw than Gary's so that seems to be a reasonable place to start. First, we need to know the time at which Alex's ball is at $y = 25.0$ m. We can use $y = y_0 + v_{0y}t + \frac{1}{2}a_y t^2$ with Alex's information to accomplish this. We know that, at this time, Gary's ball *also* has to be at $y = 25.0$ m. We can plug this time into $y = y_0 + v_{0y}t + \frac{1}{2}a_y t^2$ again—this time with Gary's information—to calculate the initial speed of Gary's ball.

SOLVE
Time it takes Alex's ball to reach $y = 25.0$ m:

$$y = y_{A0} + v_{A0y}t + \frac{1}{2}a_y t^2 = y_{A0} + v_{A0y}t + \frac{1}{2}(-g)t^2$$

$$-\frac{1}{2}gt^2 + v_{A0y}t + (y_{A0} - y) = 0$$

This is a quadratic equation, so we will use the quadratic formula to find t:

$$t = \frac{-v_{A0y} \pm \sqrt{(v_{A0y})^2 - 4\left(-\frac{g}{2}\right)(y_{A0} - y)}}{2\left(-\frac{g}{2}\right)} = \frac{-v_{A0y} \pm \sqrt{(v_{A0y})^2 + 4\left(\frac{g}{2}\right)(y_{A0} - y)}}{-g}$$

$$= \frac{-\left(-4.00\frac{m}{s}\right) \pm \sqrt{\left(-4.00\frac{m}{s}\right)^2 + 4\left(\frac{9.80\frac{m}{s^2}}{2}\right)((50.0 \text{ m}) - (25.0 \text{ m}))}}{-\left(9.80\frac{m}{s^2}\right)} = \left(\frac{4.00 \pm 22.5}{-9.80}\right) s$$

The minus sign in the numerator gives the only physical answer:

$$t = \left(\frac{4.00 - 22.5}{-9.80}\right) s = 1.89 \text{ s}$$

Speed at which Gary must launch the ball for it to be at $y = 25.0$ m at $t = 1.89$ s:

$$y = y_{G0} + v_{G0y}t + \frac{1}{2}a_y t^2 = y_{G0} + v_{G0y}t + \frac{1}{2}(-g)t^2$$

$$v_{G0y} = \frac{y - y_{G0} + \frac{1}{2}gt^2}{t} = \frac{(25.0 \text{ m}) - (1.50 \text{ m}) + \frac{1}{2}\left(9.80\frac{m}{s^2}\right)(1.89 \text{ s})^2}{1.89 \text{ s}} = \boxed{21.7\frac{m}{s}}$$

REFLECT
Gary needs to throw his ball almost 50 mph in order for it to cross paths with Alex's at $y = 25$ m! The logic behind the way we solved this problem is much more important than the algebra used to solve this problem. Be sure every step makes logical sense; the crux of the argument is that the two balls have the same position at the same time.

Get Help: Interactive Exercise – Catch the Ball

2.59

SET UP

A person falls from a height of 6 ft off of the ground. We will call the ground $y = 0$. His initial speed is zero and acceleration is $-g$ because he is undergoing free fall. Using the conversion 1 ft = 0.3048 m and $v_y^2 = v_{0y}^2 + 2a_y(y - y_0)$, we can calculate the speed at which he hits the ground.

SOLVE

Converting 6 ft into m:

$$6 \text{ ft} \times \frac{0.3048 \text{ m}}{1 \text{ ft}} = 1.83 \text{ m}$$

Solving for the final speed:

$$v_y^2 = v_{0y}^2 + 2a_y(y - y_0) = v_{0y}^2 + 2(-g)(\Delta y)$$

$$v_y = \sqrt{v_{0y}^2 - 2g(\Delta y)} = \sqrt{0^2 - 2\left(9.80\frac{\text{m}}{\text{s}^2}\right)(0 - 1.83 \text{ m})} = \boxed{6\frac{\text{m}}{\text{s}}}$$

REFLECT

A speed of 6 m/s is about 13 mph. In addition to the speed consideration, it's much easier to topple from the top step than the lower ones, hence, the warning.

2.63

SET UP

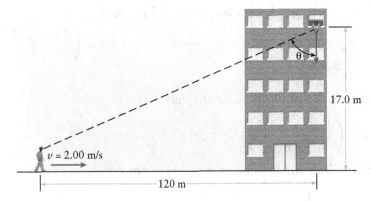

Figure 2-4 Problem 63

Mary is planning on dropping an apple out of her 17.0-m-high window to Bill. Bill is walking at a velocity of 2.00 m/s toward Mary's building and starts 120 m from directly below her window. Mary wants Bill to catch the apple, which means Bill and the apple need to be at the same location at the same time. We can use the constant acceleration equation for free fall to calculate the time it takes the apple to drop from an initial location of $y_0 = 17.0$ m to a final location of $y = 1.75$ m (presumably, Bill's height). Assume the apple has an initial velocity of zero. We can compare this to the time it takes Bill to walk 120 m in order to determine how long Mary should wait to drop the apple. Once we know how long Mary waits, we can find the distance Bill is from Mary since he is walking at a constant speed of 2.00 m/s. Bill's horizontal distance from Mary and the height Mary is in the air are two legs of a right triangle; the angle θ is related to these legs by the tangent.

SOLVE

Part a)

Time it takes Mary's apple to fall to a height of 1.75 m off the ground:

$$y = y_0 + v_{0y}t + \frac{1}{2}a_y t^2 = y_0 + 0 + \frac{1}{2}(-g)t^2$$

$$t = \sqrt{\frac{2(y - y_0)}{-g}} = \sqrt{\frac{2((1.75 \text{ m}) - (17.0 \text{ m}))}{-\left(9.80 \frac{\text{m}}{\text{s}^2}\right)}} = 1.76 \text{ s}$$

Time it takes Bill to walk 120 m:

$$v_{\text{average},x} = \frac{\Delta x}{\Delta t}$$

$$\Delta t = \frac{\Delta x}{v_{\text{average},x}} = \frac{120 \text{ m}}{2.00 \text{ m/s}} = 60.0 \text{ s}$$

Mary should wait $(60.0 \text{ s}) - (1.76 \text{ s}) = \boxed{58.2 \text{ s}}$ to drop her apple.

Part b) In 58.2 s, Bill travels $(58.2 \text{ s})\left(2.00 \frac{\text{m}}{\text{s}}\right) = 116$ m, which means he is $\boxed{4 \text{ m}}$ horizontally from the window.

Part c)

$$\tan(\theta) = \frac{4 \text{ m}}{17 \text{ m}}$$

$$\theta = \arctan\left(\frac{4}{17}\right) = \boxed{0.23 \text{ rad} = 13°}$$

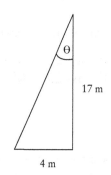

Figure 2-5 Problem 63

REFLECT

As you would expect, Bill has to be reasonably close to Mary in order for him to catch the apple. If Mary throws the apple (that is, nonzero initial velocity), Bill would need to be even closer in order to catch the apple.

Get Help: Interactive Exercise – Catch the Ball

General Problems

2.65

SET UP

Two trains are 300 m apart and traveling toward each other. Train 1 has an initial speed of 98.0 km/h and an acceleration of -3.50 m/s^2. Train 2 has an initial speed of 120 km/h and an acceleration of -4.20 m/s^2. To determine whether or not the trains collide, we can calculate the stopping distance required for each train, add them together, and see if it is

16   Chapter 2 Linear Motion

less than 300 m. If so, the trains are safe; if not, the trains crash. We know the trains' initial speeds, final speeds, and acceleration, and we are interested in finding the distance, which suggests we use $v_x^2 = v_{0x}^2 + 2a_x(x - x_0)$.

SOLVE
Stopping distance for train 1:
$$v_x^2 = v_{0x}^2 + 2a_x(x - x_0)$$

$$\Delta x = \frac{v_x^2 - v_{0x}^2}{2a_x} = \frac{0^2 - \left(98.0\frac{\text{km}}{\text{h}} \times \frac{1\text{ h}}{3600\text{ s}} \times \frac{1000\text{ m}}{1\text{ km}}\right)^2}{2\left(-3.50\frac{\text{m}}{\text{s}^2}\right)} = 106 \text{ m}$$

Stopping distance for train 2:
$$v_x^2 = v_{0x}^2 + 2a_x(x - x_0)$$

$$\Delta x = \frac{v_x^2 - v_{0x}^2}{2a_x} = \frac{0^2 - \left(120\frac{\text{km}}{\text{h}} \times \frac{1\text{ h}}{3600\text{ s}} \times \frac{1000\text{ m}}{1\text{ km}}\right)^2}{2\left(-4.20\frac{\text{m}}{\text{s}^2}\right)} = 132 \text{ m}$$

Total distance required to stop both trains: 106 m + 132 m = 238 m. This is less than the 300 m separating the trains, which means the trains will not collide . The distance separating the trains is 300 m − 238 m = 62 m .

REFLECT
As long as the trains are at least 237 m apart, they will not collide.

   Get Help: P'Cast 2.7 – Motion with Constant Acceleration I: Cleared for Takeoff!
         P'Cast 2.8 – Motion with Constant Acceleration II: Which Solution is Correct?

2.71

SET UP
A ball is dropped from an unknown height above your window. You observe that the ball takes 0.180 s to traverse the length of your window, which is 1.50 m. We can calculate the velocity of the ball when it is at the top of the window from the length of the window, the acceleration due to gravity, and the time it takes to pass by the window. Once we have the velocity at that point, we can determine the height the ball needed to fall to achieve that speed, assuming that its initial velocity was zero. Throughout the problem, we will define *down* to be negative.

SOLVE
Speed of ball at top of window:
$$y = y_0 + v_{0y}t + \frac{1}{2}a_y t^2$$

$$\Delta y = v_{0y}t + \frac{1}{2}(-g)t^2$$

$$v_{0y} = \frac{\Delta y + \frac{1}{2}gt^2}{t} = \frac{(-1.50 \text{ m}) + \frac{1}{2}\left(9.80\frac{\text{m}}{\text{s}^2}\right)(0.180 \text{ s})^2}{(0.180 \text{ s})} = -7.45\frac{\text{m}}{\text{s}}$$

Distance the ball drops to achieve a speed of 7.45 m/s:

$$v_y^2 = v_{0y}^2 + 2a_y(y - y_0)$$

$$v_y^2 = v_{0y}^2 + 2(-g)(\Delta y)$$

$$\Delta y = \frac{v_y^2 - v_{0y}^2}{2(-g)} = \frac{\left(-7.45\frac{\text{m}}{\text{s}}\right)^2 - 0}{2\left(-9.80\frac{\text{m}}{\text{s}^2}\right)} = -2.83 \text{ m}$$

The ball started at a distance of $\boxed{2.83 \text{ m}}$ above your window.

REFLECT

Be careful with the signs of Δy, a, and v in this problem.

Get Help: Interactive Exercise – Catch the Ball

2.75

SET UP

A rocket with two stages of rocket fuel is launched straight up into the air from rest. Stage 1 lasts 10.0 s and provides a net upward acceleration of 15.0 m/s². Stage 2 lasts 5.00 s and provides a net upward acceleration of 12.0 m/s². After Stage 2 finishes, the rocket continues to travel upward under the influence of gravity alone until it reaches its maximum height and then falls back toward Earth. We can split the rocket's flight into four parts: (1) Stage 1, (2) Stage 2, (3) between the end of Stage 2 and reaching the maximum height, and (4) falling from the maximum height back to Earth's surface. The acceleration of the rocket is constant over each of these legs, so we can use the constant acceleration equations to determine the total distance covered in each leg and the duration of each leg. The initial speed for legs #1 and #4 is zero, but we will need to calculate the initial speeds for legs #2 and #3. The maximum altitude is equal to the distance covered in legs #1–#3; the time required for the rocket to return to Earth is equal to the total duration of its flight, which is the sum of the durations of legs #1–#4.

SOLVE

Maximum altitude
Distance traveled in Stage 1:

$$y = y_0 + v_{0y}t + \frac{1}{2}a_yt^2$$

$$\Delta y = 0 + \frac{1}{2}\left(15.0\frac{\text{m}}{\text{s}^2}\right)(10.0 \text{ s})^2 = 750 \text{ m}$$

Speed after Stage 1:

$$v_y = v_{0y} + a_yt = 0 + \left(15.0\frac{\text{m}}{\text{s}^2}\right)(10.0 \text{ s}) = 150\frac{\text{m}}{\text{s}}$$

Distance traveled in Stage 2:

$$y = y_0 + v_{0y}t + \frac{1}{2}a_y t^2$$

$$\Delta y = \left(150\frac{m}{s}\right)(5.00\ s) + \frac{1}{2}\left(12.0\frac{m}{s^2}\right)(5.00\ s)^2 = 900\ m$$

Speed after Stage 2:

$$v_y = v_{0y} + a_y t = \left(150\frac{m}{s}\right) + \left(12.0\frac{m}{s^2}\right)(5.00\ s) = 210\frac{m}{s}$$

Distance traveled after Stage 2:

$$v_y^2 = v_{0y}^2 + 2a_y(y - y_0)$$

$$\Delta y = \frac{v_y^2 - v_{0y}^2}{2a_y} = \frac{0 - \left(210\frac{m}{s}\right)^2}{2\left(-9.80\frac{m}{s^2}\right)} = 2250\ m$$

Maximum altitude: $750\ m + 900\ m + 2250\ m = \boxed{3.90 \times 10^3\ m}$.

Time required to return to the surface
Time after Stage 2:

$$v_y = v_{0y} + a_y t$$

$$t = \frac{v_y - v_{0y}}{a_y} = \frac{0 - \left(210\frac{m}{s}\right)}{\left(-9.80\frac{m}{s^2}\right)} = 21.4\ s$$

Time from maximum height to the ground:

$$y = y_0 + v_{0y}t + \frac{1}{2}a_y t^2 = y_0 + v_{0y}t + \frac{1}{2}(-g)t^2$$

$$t = \sqrt{\frac{2(\Delta y)}{-g}} = \sqrt{\frac{2(-3.90 \times 10^3\ m)}{-\left(9.80\frac{m}{s^2}\right)}} = 28.2\ s$$

Total time of flight: $10.0\ s + 5.00\ s + 21.4\ s + 28.2\ s = \boxed{64.6\ s}$.

REFLECT
An altitude of 3900 m is around 2.5 mi. The simplest way of solving this problem was to split it up into smaller, more manageable calculations and then put all of the information together.

Get Help: Interactive Exercise – Catch the Ball

Chapter 3
Motion in Two or Three Dimensions

Conceptual Questions

3.1 Part a) No, the sum of two vectors that have different magnitudes can never equal zero. The only time two vectors can add together to have zero magnitude is when they point in opposite directions and have the same magnitude.

Part b) Yes, the sum of three (or more) vectors with different magnitudes can be equal to zero.

Get Help: Picture It – Adding and Subtracting Vectors

3.5 Air resistance can be modeled as a force that depends on the velocity. The greater the speed, the greater the magnitude of the drag force. Also, the drag force points in the direction opposite to the direction of motion. Therefore, when the effects of air resistance are taken into account, the projectile's speed in the vertical direction will decrease more rapidly as it ascends and increase less rapidly when it comes down compared to when the effects of air resistance are ignored. When effects of air resistance are ignored, a projectile experiences no acceleration in the horizontal direction and, therefore, travels at constant speed horizontally. Due to the drag force resulting from air resistance, however, a projectile experiences a horizontal acceleration that causes its horizontal speed to decrease.

3.9 Part a) Maximum range is achieved at a launch angle of 45°.

Part b) The longest time of flight is achieved by launching the projectile at an angle of 90° (that is, straight up). We can see this from rearranging kinematic equation 3-12b:
$$t_{peak} = \frac{v_0 \sin(\theta)}{g}.$$
Part c) The greatest height is achieved by launching the projectile at an angle of 90° (that is, straight up). We can see this from rearranging the kinematic equation 3-13b:
$$(y - y_0)_{peak} = \frac{1}{2} \frac{v_0^2 \sin^2(\theta)}{g}.$$

3.13 Yes, the ape is accelerating at the bottom of its swing. The force from the vine is accelerating the ape upward since it is at the bottom of the swing that it changes from traveling downward to traveling upward.

Multiple-Choice Questions

3.17 **B (45°).** $\theta = \tan^{-1}\left(\frac{A_y}{A_x}\right) = \tan^{-1}(1) = 45°.$

3.21 D ($d_2 = 4d_1$). Equations 3-13a and 3-13b can be rearranged to solve for the horizontal range equation:

$$d_1 = \frac{2v_{0x}^2 \sin(\theta)\cos(\theta)}{g}$$

$$d_2 = \frac{2(2v_{0x})^2 \sin(\theta)\cos(\theta)}{g} = 4\frac{2v_{0x}^2 \sin(\theta)\cos(\theta)}{g} = 4d_1$$

Get Help: P'Cast 3.6 – How High Does It Go?
P'Cast 3.7 – How Long Is It In Flight?

Estimation/Numerical Analysis

3.29

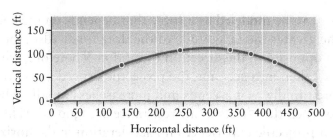

Figure 3-1 Problem 29

The ball reached a maximum height of $\Delta y = 110$ ft:

$$v_y^2 = v_{0y}^2 + 2a_y(y - y_0)$$

$$v_{0y} = \sqrt{-2a_y(\Delta y)} = \sqrt{-2\left(-32\frac{\text{ft}}{\text{s}^2}\right)(110 \text{ ft})} = 83.9\frac{\text{ft}}{\text{s}}$$

Estimating the initial launch angle of the ball from the graph to be 30°:

$$v_{0y} = v_0 \sin(30°) = \frac{v_0}{2}$$

$$v_0 = 2v_{0y} = 2\left(83.9\frac{\text{ft}}{\text{s}}\right) = 170\frac{\text{ft}}{\text{s}}$$

But the air resistance is not completely negligible here. How much is there? The peak occurs at a horizontal distance of 300 ft and the ball hits Earth at around 540 ft. Without air resistance, the ball should have landed at around 800 ft, so 30% of the forward progress is lost. Therefore, a reasonable estimate of the initial speed is (1.3)(170 ft/s) = 220 ft/s.

Get Help: Interactive Exercise – Balloon
Interactive Exercise – Arrow
P'Cast 3.6 – How High Does It Go?
P'Cast 3.7 – How Long Is It In Flight?

Problems

3.33

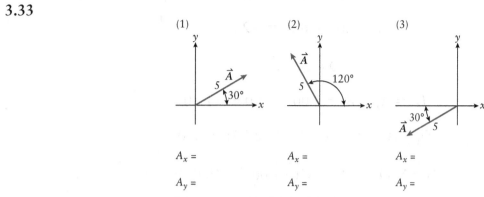

Figure 3-2 Problem 33

SET UP

A vector has a magnitude of 5 and is shown in three different orientations. Since we are given the angles in each case, we can calculate the components directly using $A_x = A\cos(\theta)$ and $A_y = A\sin(\theta)$.

SOLVE

(1)
$$A_x = 5\cos(30°) = \boxed{4.3}$$
$$A_y = 5\sin(30°) = \boxed{2.5}$$

(2)
$$A_x = 5\cos(120°) = -5\cos(60°) = \boxed{-2.5}$$
$$A_y = 5\sin(120°) = 5\sin(60°) = \boxed{4.3}$$

(3)
$$A_x = 5\cos(210°) = -5\cos(30°) = \boxed{-4.3}$$
$$A_y = 5\sin(210°) = -5\sin(30°) = \boxed{-2.5}$$

REFLECT

It is a good idea to redraw the right triangle each time to ensure you are taking the sine or cosine of the correct angle.

Get Help: Interactive Exercise – Vector Subtraction
P'Cast 3.1 – Different Descriptions of a Vector
P'Cast 3.3 – Adding Vectors Using Components

3.37

SET UP

Two vectors are given. We can calculate their difference by subtracting their components and then determining the magnitude and angle using the Pythagorean theorem and the tangent, respectively.

SOLVE

$$A_x = 2.00 \text{ and } A_y = 6.00$$

$$B_x = 3.00 \text{ and } B_y = -2.00$$

$$\vec{D} = \vec{A} - \vec{B}$$

$$D_x = A_x - B_x = 2.00 - 3.00 = -1.00$$

$$D_y = A_y - B_y = 6.00 - (-2.00) = 8.00$$

$$D = \sqrt{D_x^2 + D_y^2} = \sqrt{(-1.00)^2 + (8.00)^2}$$

$$\boxed{D = 8.06}$$

$$\theta = \tan^{-1}\left(\frac{8.00}{-1.00}\right) = \boxed{1.70 \text{ rad}}$$

REFLECT

An angle of 1.70 radians is equal to 97.1°. The difference $\vec{A} - \vec{B}$ is the same as the sum $\vec{A} + (-1)\vec{B}$.

Get Help: Interactive Exercise – Vector Subtraction
P'Cast 3.1 – Different Descriptions of a Vector
P'Cast 3.3 – Adding Vectors Using Components

3.41

SET UP

The initial and final velocity vectors of an object are given. We can calculate the components of the average acceleration vector of the object by calculating the changes in the components of the velocity vector and dividing them by the total time of 5 s. We can use the components of the average acceleration vector to determine its magnitude. We are not given the intermediate velocity vectors, so we cannot determine if the acceleration was uniform.

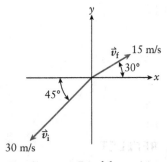

Figure 3-3 Problem 41

SOLVE

$$v_{ix} = -\left(30\frac{\text{m}}{\text{s}}\right)\cos(45°) = -21\frac{\text{m}}{\text{s}}$$

$$v_{iy} = -\left(30\frac{\text{m}}{\text{s}}\right)\sin(45°) = -21\frac{\text{m}}{\text{s}}$$

$$v_{fx} = \left(15\frac{\text{m}}{\text{s}}\right)\cos(30°) = 13\frac{\text{m}}{\text{s}}$$

$$v_{fy} = \left(15\frac{\text{m}}{\text{s}}\right)\sin(30°) = 7.5\frac{\text{m}}{\text{s}}$$

$$a_x = \frac{\Delta v_x}{\Delta t} = \frac{v_{fx} - v_{ix}}{t_2 - t_1} = \frac{\left(13\frac{m}{s}\right) - \left(-21\frac{m}{s}\right)}{(5\text{ s}) - (0\text{ s})} = 6.8\frac{m}{s^2}$$

$$a_y = \frac{\Delta v_y}{\Delta t} = \frac{v_{fy} - v_{iy}}{t_2 - t_1} = \frac{\left(7.5\frac{m}{s}\right) - \left(-21\frac{m}{s}\right)}{(5\text{ s}) - (0\text{ s})} = 5.7\frac{m}{s^2}$$

$$a = \sqrt{a_x^2 + a_y^2} = \sqrt{\left(6.8\frac{m}{s^2}\right)^2 + \left(5.7\frac{m}{s^2}\right)^2} = \boxed{9\frac{m}{s^2}}$$

No, it's not possible to know whether the acceleration was uniform from the information given. We would need to know the acceleration as a function of time in order to answer that question.

REFLECT

The average acceleration vector points in the same direction as the change in the velocity. Its magnitude is scaled by the time interval.

3.47

SET UP

Five balls are thrown off the edge of a cliff with the same speed but at different angles. We can use intuition and the constant acceleration equations to determine which ball travels the farthest horizontal distance, which ball takes the longest to hit the ground, and which ball has the greatest speed when landing.

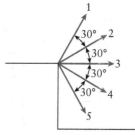

Figure 3-4 Problem 47

SOLVE

Part a) Balls 1 and 2 should travel the farthest since they are thrown upward. Balls 4 and 5 should travel the shortest distance since they are thrown downward. The initial x velocity for ball 2 is larger than ball 1, so it should travel farther horizontally before hitting the ground. Therefore, ball 2 travels the farthest followed by balls 1, 3, 4, and 5.

Part b) Ball 1 should take the longest to hit the ground because it has the largest, positive initial y velocity, followed by balls 2, 3, 4, and 5.

Part c)

$$v_y^2 = v_{0y}^2 + 2a_y(y - y_0)$$
$$v_y = \sqrt{v_{0y}^2 + 2a_y(\Delta y)}$$

The initial speed, the acceleration, and the vertical displacement traveled are all the same for the five cases. Therefore, all five balls will land with the same vertical component of velocity.

REFLECT

Be careful when searching through the text for equations to use. Some of them will only work for specific cases and not in general.

Get Help: Interactive Exercise – Balloon
Interactive Exercise – Arrow
P'Cast 3.6 – How High Does It Go?
P'Cast 3.7 – How Long Is It In Flight

3.49

SET UP

A tiger leaps out of a tree that is 4.00 m tall. He lands 5.00 m from the base of the tree. The tiger's initial velocity is completely in the horizontal direction, which means the y component of his initial velocity is zero. We will choose the tiger's starting point to be the origin and up to be positive y. We can assume the tiger is jumping in the positive x direction. We can use the y component information to solve for t in terms of known quantities and then solve for the initial speed.

SOLVE

Know/Don't Know table:

	x		y
x_0	0 m	y_0	0 m
x_f	5.00 m	y_f	−4.00 m
v_{0x}	WANT	v_{0y}	0
v_x	?	v_y	?
a_x	0	a_y	−9.80 m/s²
t	?	t	?

Solving for initial speed:

$$y = y_0 + v_{0y}t + \frac{1}{2}a_y t^2 = y_0 + 0 + \frac{1}{2}a_y t^2$$

$$t = \sqrt{\frac{2(\Delta y)}{a_y}}$$

$$x = x_0 + v_{0x}t + \frac{1}{2}a_x t^2$$

$$\Delta x = v_{0x}t + 0 = v_{0x}\left(\sqrt{\frac{2(\Delta y)}{a_y}}\right)$$

$$v_{0x} = \Delta x\left(\sqrt{\frac{a_y}{2(\Delta y)}}\right) = (5.00 \text{ m})\left(\sqrt{\frac{-9.80\frac{\text{m}}{\text{s}^2}}{2(-4.00 \text{ m})}}\right) = \boxed{5.53\frac{\text{m}}{\text{s}}}$$

REFLECT

We could have chosen the base of the tree to be the origin rather than the tiger's initial location. This would not have affected the sign or magnitude of the y displacement.

Get Help: Interactive Exercise – Balloon
Interactive Exercise – Arrow
P'Cast 3.6 – How High Does It Go?
P'Cast 3.7 – How Long Is It In Flight?

3.53

SET UP

A ball is attached to a 0.870-m-long string and is moving in a circle with a constant speed of 3.36 m/s. The ball is undergoing uniform circular motion, so its acceleration has a magnitude of $a_{cent} = \dfrac{v^2}{r}$ and points toward the center of the circle.

SOLVE

$$a_{cent} = \dfrac{v^2}{r} = \dfrac{\left(3.36\dfrac{m}{s}\right)^2}{0.870 \text{ m}} = \boxed{13.0\dfrac{m}{s^2} \text{ inward}}$$

REFLECT

Centripetal acceleration always points toward the center of the circle.

Get Help: P'Cast 3.9 – Around the Bend

3.57

SET UP

A Ferris wheel is 76 m in diameter and completes a revolution in 20 min, which means a point on the rim will travel a length of one circumference in 20 min. We will assume the Ferris wheel is undergoing uniform circular motion, so its acceleration is equal to $a_{cent} = \dfrac{v^2}{r}$.

SOLVE

$$v = \dfrac{1 \text{ rev}}{20 \text{ min}} \times \dfrac{1 \text{ min}}{60 \text{ s}} \times \dfrac{\pi(76 \text{ m})}{1 \text{ rev}} = 0.199\dfrac{m}{s}$$

$$a_{cent} = \dfrac{v^2}{r} = \dfrac{\left(0.199\dfrac{m}{s}\right)^2}{\left(\dfrac{76 \text{ m}}{2}\right)} = \boxed{1 \times 10^{-3}\dfrac{m}{s^2}}$$

REFLECT

We are given the *diameter* of the wheel, not the *radius*. The radial acceleration uses the radius, so be sure to divide by 2 when performing the calculation.

Get Help: P'Cast 3.9 – Around the Bend
P'Cast 3.10 – Orbital Speed

3.61

SET UP

The space shuttle is in an orbit about 300 km above the surface of Earth, which has a radius of 6.38×10^6 m. We will assume the orbit is circular and that the space shuttle is moving with a constant speed. The shuttle sweeps out a circle of radius $r = r_{Earth} + r_{orbit}$ and covers a distance of $2\pi r$ in one period T. The acceleration of the shuttle is $a_{cent} = \dfrac{v^2}{r}$ and points toward the center of Earth.

SOLVE

$$v = \frac{2\pi(r_{Earth} + r_{orbit})}{T} = \frac{2\pi((6.38 \times 10^6 \text{ m}) + (3.00 \times 10^5 \text{ m}))}{(5.43 \times 10^3 \text{ s})} = 7730 \frac{\text{m}}{\text{s}}$$

$$a_{cent} = \frac{v^2}{r_{Earth} + r_{orbit}} = \frac{\left(7730 \frac{\text{m}}{\text{s}}\right)^2}{(6.38 \times 10^6 \text{ m}) + (3.00 \times 10^5 \text{ m})}$$

$$= \boxed{9 \frac{\text{m}}{\text{s}^2} \text{ toward the center of Earth}}$$

REFLECT

This is about 90% of the acceleration due to gravity at Earth's surface.

Get Help: P'Cast 3.9 – Around the Bend
P'Cast 3.10 – Orbital Speed

3.63

SET UP

A peregrine falcon pulls out of a dive and into a circular arc at a speed of 100 m/s. It experiences an acceleration of $0.6g$, where $g = 9.80$ m/s^2. Assuming the bird's speed is constant, we can calculate the radius of the turn from the centripetal acceleration.

SOLVE

$$a_{cent} = \frac{v^2}{r}$$

$$r = \frac{v^2}{a_{cent}} = \frac{\left(100 \frac{\text{m}}{\text{s}}\right)^2}{(0.6)\left(9.80 \frac{\text{m}}{\text{s}^2}\right)} = 1700 \text{ m} = \boxed{2 \text{ km}}$$

REFLECT

Peregrine falcons are known to be one of the fastest animals on Earth. A speed of 100 m/s is roughly 220 mph!

Get Help: P'Cast 3.9 – Around the Bend
P'Cast 3.10 – Orbital Speed

General Problems

3.67

SET UP

Steve wants to throw a football to Jerry at an initial speed of 15.0 m/s at an angle of 45°. Jerry runs past Steve at a speed of 8.00 m/s in the direction the football will be thrown. In order for Jerry to catch the football, Jerry and the football need to be at the same place at the same time. We can calculate the distance the football will travel and the time it takes for it to travel that distance. Because we are told that the football is caught at the same height from which it is released, we can derive an equation for the maximum horizontal range using the kinematic equations. We can compare this time with the time it takes Jerry to run the same distance. The difference between these times is how long Steve should wait to throw the ball.

In part (b), Steve starts to run with the football at a speed of 1.50 m/s down the field as well. This means we need to use the speeds for the football and for Jerry relative to Steve. The football will still cover 23.0 m since its speed *relative to the ground* is still 15.0 m/s.

SOLVE

Part a)

Total time the football is in the air:

$$v_{0y} = v_0 \sin \theta$$

$$y = y_0 + v_{0y}t + \frac{1}{2}a_y t^2$$

$$0 = 0 + (v_0 \sin \theta)t + \frac{1}{2}(-g)t^2$$

$$t = \frac{2v_0 \sin \theta}{g}$$

Maximum horizontal distance the football travels in this time:

$$v_{0x} = v_0 \cos \theta$$

$$x = x_0 + v_{0x}t$$

$$x - x_0 = (v_0 \cos \theta)\left(\frac{2v_0 \sin \theta}{g}\right)$$

Distance the football travels:

$$(x - x_0)_{max} = \frac{2v_0^2 \sin(\theta)\cos(\theta)}{g} = \frac{2\left(15.0 \frac{m}{s}\right)^2 \sin(45°)\cos(45°)}{9.80 \frac{m}{s^2}} = 23.0 \text{ m}$$

Time it takes the football to travel 23.0 m:

$$x = x_0 + v_{0x}t = x_0 + v_0 \cos(45°)t$$

$$t = \frac{\Delta x}{v_0 \cos(45°)} = \frac{23.0 \text{ m}}{\left(15.0 \frac{\text{m}}{\text{s}}\right)\cos(45°)} = 2.17 \text{ s}$$

Time it takes Jerry to run 23.0 m:

$$v_{Jx} = \frac{\Delta x}{\Delta t}$$

$$\Delta t = \frac{\Delta x}{v_{Jx}} = \frac{23.0 \text{ m}}{8.00 \frac{\text{m}}{\text{s}}} = 2.88 \text{ s}$$

Steve should wait 2.88 s − 2.17 s = $\boxed{0.71 \text{ s}}$ to throw the football.

Part b)

Time it takes the football to travel 23.0 m at a relative speed of 13.5 m/s:

$$x = x_0 + v_{0x}t = x_0 + v_0 \cos(45°)t$$

$$t = \frac{\Delta x}{v_0 \cos(45°)} = \frac{23.0 \text{ m}}{\left(13.5 \frac{\text{m}}{\text{s}}\right)\cos(45°)} = 2.41 \text{ s}$$

Time it takes Jerry to run 23.0 m at a relative speed of 6.5 m/s:

$$v_{Jx} = \frac{\Delta x}{\Delta t}$$

$$\Delta t = \frac{\Delta x}{v_{Jx}} = \frac{23.0 \text{ m}}{6.50 \frac{\text{m}}{\text{s}}} = 3.54 \text{ s}$$

Steve should wait 3.54 s − 2.41 s = $\boxed{1.13 \text{ s}}$ to throw the football.

REFLECT

It makes sense that Steve should wait longer to release the football when he is running rather than standing still.

Get Help: Interactive Exercise – Balloon
Interactive Exercise – Arrow
P'Cast 3.6 – How High Does It Go?
P'Cast 3.7 – How Long Is It In Flight?

3.73

SET UP

A water balloon is thrown horizontally at an initial speed of 2.00 m/s from the top of a 6.00-m-tall building. At the same time, a second water balloon is thrown straight down from the same height at an initial speed of 2.00 m/s. To determine which balloon hits the ground first, we need to calculate the time it takes each balloon to hit the ground; whichever balloon has the shorter time will hit the ground first. Since the balloons start with the same initial speed and fall the same distance, they will hit the ground with the same final speed.

SOLVE
Know/Don't Know table for balloon 1:

	x		y
x_0	0 m	y_0	6.00 m
x_f	?	y_f	0
v_{0x}	2.00 m/s	v_{0y}	0
v_x	?	v_y	?
a_x	0	a_y	−9.80 m/s²
t	?	t	?

Time for balloon 1 to hit the ground:

$$y = y_0 + v_{0y}t + \frac{1}{2}a_y t^2 = y_0 + 0 + \frac{1}{2}a_y t^2$$

$$t = \sqrt{\frac{2(\Delta y)}{a_y}} = \sqrt{\frac{2(-6.00 \text{ m})}{\left(-9.80 \frac{\text{m}}{\text{s}^2}\right)}} = 1.11 \text{ s}$$

Know/Don't Know table for balloon 2:

	y
y_0	6.00 m
y_f	0
v_{0y}	−2.00 m/s
v_y	?
a_y	−9.80 m/s²
t	?

Time for balloon 2 to hit the ground:

$$\frac{1}{2}a_y t^2 + v_{0y}t - \Delta y = 0$$

$$t = \frac{-v_{0y} \pm \sqrt{v_{0y}^2 - 4\left(\frac{1}{2}a_y\right)(-\Delta y)}}{2\left(\frac{1}{2}a_y\right)} = \frac{-\left(-2.00\frac{\text{m}}{\text{s}}\right) \pm \sqrt{\left(-2.00\frac{\text{m}}{\text{s}}\right)^2 + 2\left(-9.80\frac{\text{m}}{\text{s}^2}\right)(-6.00 \text{ m})}}{\left(-9.80\frac{\text{m}}{\text{s}^2}\right)}$$

$$= \frac{-2.00 \pm 11.0}{-9.80} \text{ s}$$

Taking the negative root: $t = 0.918$ s.

Balloon 2 lands 0.19 s before balloon 1.

Final speed of the balloons:

$$v_y^2 = v_{0y}^2 + 2a_y(y - y_0)$$

$$v_y = \sqrt{v_{0y}^2 + 2a_y(\Delta y)}$$

Since the balloons start with the same initial speed and fall the same distance, the balloons will land with the same speed.

REFLECT
It will save you time and frustration in the long run if you solve each problem algebraically first and then plug in numbers at the end. For example, we could easily see the balloons have the same final speed from the general, algebraic answer in one calculation, rather than two.

Get Help: Interactive Exercise – Balloon
Interactive Exercise – Arrow
P'Cast 3.6 – How High Does It Go?
P'Cast 3.7 – How Long Is It In Flight?

3.79

SET UP
Gabriele Reinsch threw a discus 76.80 m. She launched it an angle of 45° from a height of 2.0 m above the ground. We will take Gabriele's feet to be the origin, the positive x direction to be the direction she threw the discus, and up to be the positive y direction. We can use the x information to solve for t and plug this into the y equation to solve for her initial speed. Remember that $\cos(45°) = \sin(45°) = \frac{1}{\sqrt{2}}$; this will help simplify the algebra.

SOLVE
Know/Don't Know table:

	x		y
x_0	0 m	y_0	2.0 m
x_f	76.80 m	y_f	0 m
v_{0x}	$v_0\cos(45°)$	v_{0y}	$v_0\sin(45°)$
v_x	?	v_y	?
a_x	0	a_y	−9.80 m/s²
t	?	t	?

$$x = x_0 + v_{0x}t + \frac{1}{2}a_xt^2 = x_0 + v_0\cos(45°)t + 0$$

$$t = \frac{\Delta x}{v_0\cos(45°)}$$

$$y = y_0 + v_{0y}t + \frac{1}{2}a_yt^2 = y_0 + v_0\sin(45°)\left(\frac{\Delta x}{v_0\cos(45°)}\right) + \frac{1}{2}a_y\left(\frac{\Delta x}{v_0\cos(45°)}\right)^2 = y_0 + \Delta x + a_y\left(\frac{\Delta x}{v_0}\right)^2$$

$$v_0 = \sqrt{\frac{-a_y(\Delta x)^2}{(\Delta x) - (\Delta y)}} = \sqrt{\frac{-\left(-9.80\frac{m}{s^2}\right)(76.80 \text{ m})^2}{(76.80 \text{ m}) - (-2.0 \text{ m})}} = 27.1 \frac{m}{s}$$

REFLECT

We cannot use the maximum horizontal range formula from the text because the discus does not land at the same height it was released.

Get Help: Interactive Exercise – Balloon
Interactive Exercise – Arrow
P'Cast 3.6 – How High Does It Go?
P'Cast 3.7 – How Long Is It In Flight?

3.83

SET UP

A sample in a centrifuge initially experiences a radial acceleration of $a_1 = 100g$. The acceleration then increases by a factor of 8, $a_2 = 800g$. We can determine the new rotation speed in terms of the original rotation speed by looking at the ratio of the two radial accelerations. The radius is the same in both cases, since the size of the centrifuge doesn't change.

SOLVE

$$a_{\text{cent, 1}} = 100g = \frac{v_1^2}{r}$$

$$a_{\text{cent, 2}} = 800g = \frac{v_2^2}{r}$$

$$\frac{a_{\text{cent, 2}}}{a_{\text{cent, 1}}} = \frac{800g}{100g} = 8 = \frac{\left(\frac{v_1^2}{r}\right)}{\left(\frac{v_2^2}{r}\right)} = \frac{v_1^2}{v_2^2}$$

$$\boxed{v_2 = v_1\sqrt{8} \approx 2.83 v_1}$$

REFLECT

If a problem asks you to find the factor by which a quantity increases or decreases, first try taking a ratio of the quantities. Usually some terms will remain constant over the process and will cancel out in your ratio.

Get Help: P'Cast 3.9 – Around the Bend
P'Cast 3.10 – Orbital Speed

Chapter 4
Forces and Motion I: Newton's Laws

Conceptual Questions

4.1 Yes, the direction of the net force will determine the direction of the acceleration vector. Mass is a scalar quantity and will not affect the direction.

4.7 You can double the rope up. Each of the two strands pulls only 425 N.

        Get Help: Interactive Exercise – Tension
                Interactive Exercise – Two Strings

4.9 Yes, both the Sun and Earth pull each other and make up a force pair associated with the gravitational interaction.

4.15 The net force on the bathroom scale is zero since it is at rest.

4.19 It's easier to lift a truck on the Moon. Weight is due to gravity. The Moon has less gravity than Earth.

4.23 A bird begins flying by pushing with its wings backward against the air. The air then exerts an equal and opposite force forward on the bird, and it is this force on the bird that propels it forward.

Multiple-Choice Questions

4.25 B (rear). The car accelerates forward quickly due to the rear-end collision, while the person stays in place.

4.31 D (at rest or in motion with a constant velocity). A net force of zero means the acceleration is also zero.

Estimation/Numerical Analysis

4.37 A baseball throw can be in the range of 40 m/s (about 89 mph). A baseball has a mass of around 0.15 kg. The powerful part of the throw lasts about 0.05 s. This gives

$$F_{ext,x} = m\frac{\Delta v_x}{\Delta t} = (0.15 \text{ kg})\frac{\left(40\frac{\text{m}}{\text{s}}\right)}{0.05 \text{ s}} \approx 120 \text{ N}.$$

        Get Help: P'Cast 4.1 – Small But Forceful

4.45 The weight of the rider will reach its maximum value during the ascent from the ground floor to the fourth floor. This is because the weight of the rider will be added to the force required to accelerate from 0 m/s to 10 m/s.

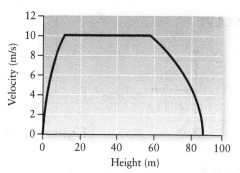

Figure 4-1 Problem 45

Problems

4.49

SET UP

The engine of a 1250-kg car can deliver a maximum force of 15,000 N. Since this is the only force acting on the car, we can set this force equal to the mass times the acceleration and solve for the magnitude of the acceleration.

SOLVE

$$\sum F_{ext,x} = ma_x$$

$$a_x = \frac{\sum F_{ext,x}}{m} = \frac{15{,}000 \text{ N}}{1250 \text{ kg}} = \boxed{12 \frac{\text{m}}{\text{s}^2}}$$

REFLECT

Because there is only one force acting on the car, this force is equal to the net force.

Get Help: P'Cast 4.1 – Small But Forceful

4.53

SET UP

A tuna has a mass of 250 kg. Its weight is equal to the magnitude of the force of gravity on it, which is equal to the fish's mass multiplied by the acceleration due to gravity.

SOLVE

$$w_{tuna,y} = m_{tuna}g = (250 \text{ kg})\left(9.80 \frac{\text{m}}{\text{s}^2}\right) = \boxed{2.5 \times 10^3 \text{ N}}$$

REFLECT

This is a weight of 550 lb.

Get Help: P'Cast 4.3 – Let Sleeping Cats Lie

4.57

SET UP

A person is pushing a box across a smooth floor at a steadily increasing speed, which means the box is accelerating along that direction. The force due to the man on the box acts horizontally, while the normal force and the force due to gravity act vertically.

SOLVE

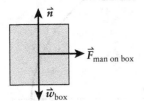

Figure 4-2 Problem 57

REFLECT
The normal force and force due to gravity are equal in magnitude because the box is not accelerating in the vertical direction.

4.61

SET UP
Three rugby players are pulling horizontally on ropes attached to a stationary box. Player 1 pulls with a force of 100.0 N at an angle of −60.0°. Player 2 pulls with a force of 200.0 N at an angle of +37.0°. We can use Newton's second law to calculate the components of Player 3's force. Since the box is at rest, the acceleration is equal to zero. Once we know the components of Player 3's force, we can calculate its magnitude and direction and draw it on the figure. Player 3's rope breaks, which means $F_3 = 0$ N, and Player 2 compensates for this by pulling with a force of only 150.0 N. The direction of the box's resulting acceleration is equal to the direction of the net force due to Players 1 and 2. We can calculate the mass of the box from Newton's second law.

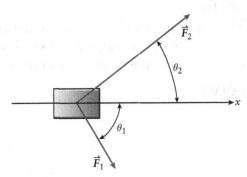

Figure 4-3 Problem 61

SOLVE
Part a)

$$\sum F_{ext,x} = F_{1x} + F_{2x} + F_{3x} = F_1\cos(\theta_1) + F_2\cos(\theta_2) + F_{3x} = m_{box}a_x = 0$$

$$F_{3x} = -F_1\cos(\theta_1) - F_2\cos(\theta_2) = -(100.0\text{ N})\cos(60.0°) - (200.0\text{ N})\cos(37.0°) = \boxed{-2.10 \times 10^2 \text{ N}}$$

$$\sum F_{ext,y} = -F_{1y} + F_{2y} + F_{3y} = -F_1\sin(\theta_1) + F_2\sin(\theta_2) + F_{3y} = m_{box}a_y = 0$$

$$F_{3y} = F_1\sin(\theta_1) - F_2\sin(\theta_2) = (100.0\text{ N})\sin(60.0°) - (200.0\text{ N})\sin(37.0°) = \boxed{-33.8 \text{ N}}$$

Part b)

$$F_3 = \sqrt{F_{3x}^2 + F_{3y}^2} = \sqrt{(-2.10 \times 10^2 \text{ N})^2 + (-33.8\text{ N})^2} = 213 \text{ N}$$

$$\theta_3 = \tan^{-1}\left(\frac{F_{3y}}{F_{3x}}\right) = \tan^{-1}\left(\frac{-33.8\text{ N}}{-2.10 \times 10^2 \text{ N}}\right) = 189.1°$$

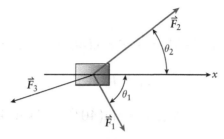

Figure 4-4 Problem 61

Part c)

$$\sum F_{ext,x} = F_{1x} + F_{2x} = F_1 \cos(\theta_1) + F_2 \cos(\theta_2) = m_{box} a_x$$

$$F_1 \cos(\theta_1) + F_2 \cos(\theta_2) = (100.0 \text{ N}) \cos(60.0°) + (150.0 \text{ N}) \cos(37.0°) = 1.70 \times 10^2 \text{ N} = m_{box} a_x$$

$$\sum F_{ext,y} = -F_{1y} + F_{2y} = -F_1 \sin(\theta_1) + F_2 \sin(\theta_2) = m_{box} a_y$$

$$-F_1 \sin(\theta_1) + F_2 \sin(\theta_2) = -(100.0 \text{ N}) \sin(60.0°) + (150.0 \text{ N}) \sin(37.0°) = 3.67 \text{ N} = m_{box} a_y$$

$$\theta = \tan^{-1}\left(\frac{3.67 \text{ N}}{1.70 \times 10^2 \text{ N}}\right) = \boxed{1.24°}$$

Part d)

$$a_x = a \cos(\theta) = \left(10.0 \frac{\text{m}}{\text{s}^2}\right) \cos(1.24°) = 10.0 \frac{\text{m}}{\text{s}^2}$$

$$m_{box} = \frac{1.70 \times 10^2 \text{ N}}{a_x} = \frac{1.70 \times 10^2 \text{ N}}{10.0 \frac{\text{m}}{\text{s}^2}} = \boxed{17.0 \text{ kg}}$$

REFLECT

Players 1 and 2 pull toward $+x$, so Player 3 must pull toward $-x$ if the box is to remain stationary. If Player 3 stops pulling, the box must accelerate toward $+x$.

4.65

SET UP

Two ropes support a 100.0-kg streetlight. One rope pulls up and to the left at an angle of 40° above the horizontal with a tension T_1. The other rope pulls up and to the right at an angle of 40° above the horizontal with a tension T_2. The forces acting on the streetlight are the tension in the first rope, the tension in the second rope, and the force of gravity pointing straight down. The streetlight is at rest, which means its acceleration is equal to zero in all directions. We can solve for the tension in each rope by solving Newton's second law in component form. The tension in each rope is equal to the magnitude T_1 or T_2.

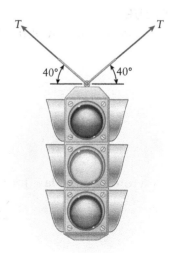

Figure 4-5 Problem 65

SOLVE

$$\sum F_{ext,x} = T_{1x} + T_{2x} = -T_1\cos(40°) + T_2\cos(40°) = 0$$

$$T_1 = T_2 \equiv T$$

$$\sum F_{ext,y} = T_{1y} + T_{2y} + w_{streetlight} = T_1\sin(40°) + T_2\sin(40°) - m_{streetlight}g = 0$$

$$2T\sin(40°) = m_{streetlight}g$$

$$T = \frac{m_{streetlight}g}{2\sin(40°)} = \frac{(100.0 \text{ kg})\left(9.80\frac{\text{m}}{\text{s}^2}\right)}{2\sin(40°)} = \boxed{762 \text{ N}}$$

REFLECT
Because the ropes are symmetric, they must have the same tension.

Get Help: Interactive Exercise – Tension
Interactive Exercise – Two Strings
P'Cast 4.4 – What's the Angle?

4.69

SET UP
A train is made up of a locomotive and 10 identical freight cars, each of mass M. The magnitude of the train's acceleration is 2 m/s². The magnitude of the force between the locomotive and the first car is 100,000 N. If we treat all 10 freight cars as the "train," the force between the locomotive and the "train" is equal to the force between the locomotive and the first car. We can draw a free-body diagram of the "train" in order to calculate the mass of one of the cars. We can then draw the free-body diagram of the tenth car and apply Newton's second law in order to calculate the magnitude of the force between the ninth and tenth cars. We'll define positive x as pointing to the right in the above figure.

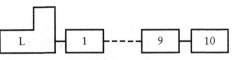

Figure 4-6 Problem 69

SOLVE
Free-body diagram of the "train":

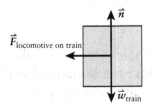

Figure 4-7 Problem 69

Newton's second law for the "train":

$$\sum F_{ext \text{ on train},x} = -F_{locomotive \text{ on train},x} = (10M)a_x$$

$$M = \frac{-F_{locomotive \text{ on train},x}}{10a_x}$$

Free-body diagram of car 10:

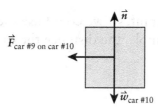

Figure 4-8 Problem 69

Newton's second law for car 10:

$$\sum F_{\text{ext on car 10},x} = -F_{\text{car 9 on car 10},x} = Ma_x = \left(\frac{-F_{\text{locomotive on train},x}}{10a_x}\right)a_x$$

$$F_{\text{car 9 on car 10},x} = \frac{F_{\text{locomotive on train},x}}{10} = \frac{100{,}000 \text{ N}}{10} = \boxed{10{,}000 \text{ N}}$$

REFLECT

Rather than treating all 10 cars as one object called the "train," we could have drawn 10 different free-body diagrams—one for each freight car. The force of car 1 on car 2 is equal in magnitude and opposite in direction to the force of car 2 on car 1. Applying Newton's second and third laws for each car in the series will give you the same answer as above.

4.73

SET UP

A 60.0-kg person is riding in a car. While the car uniformly accelerates from 0 to 28.0 m/s, the person feels a horizontal force of 400 N; this is the only force acting on the person in the horizontal direction. We can use Newton's second law to calculate the acceleration and then use the definition of constant acceleration to calculate the time it takes the car to reach 28.0 m/s.

SOLVE

Newton's second law for the person:

$$\sum F_{\text{ext},x} = F_{\text{car on person}} = m_{\text{person}} a_x$$

$$a_x = \frac{F_{\text{car on person}}}{m_{\text{person}}}$$

Calculating the time:

$$a_x = \frac{\Delta v_x}{\Delta t}$$

$$\Delta t = \frac{\Delta v_x}{a_x} = \frac{\Delta v_x}{\left(\dfrac{F_{\text{car on person}}}{m_{\text{person}}}\right)} = \frac{m_{\text{person}} \Delta v_x}{F_{\text{car on person}}} = \frac{(60.0 \text{ kg})\left(\left(28.0\dfrac{\text{m}}{\text{s}}\right) - 0\right)}{400 \text{ N}} = \boxed{4 \text{ s}}$$

REFLECT

The acceleration of the car is constant, so we could use the constant acceleration equations to calculate other quantities, for example, the distance the car travels.

4.77

SET UP

Two blocks, resting on two different inclined planes, are attached by a string. The block on the left has a mass $M_1 = 6.00$ kg and rests on an incline at $60.0°$. The block on the right has a mass M_2 and rests on an incline at $25°$. We need to find the value of M_2 such that both blocks remain at rest. We'll use two different but related coordinate systems for the two blocks. For both blocks the axes will be parallel and perpendicular to the inclined plane. For block M_1 up the ramp and out of the ramp are positive, while down the ramp and out of the ramp are positive for M_2. We can draw free-body diagrams and apply Newton's second law in component form for both blocks. The tension acting on each block will be identical in magnitude. We are only interested in the forces and acceleration acting parallel to the planes because the blocks do not leave the plane; the acceleration in the perpendicular direction is always zero. Both blocks are at rest, which means the acceleration of each block is zero and we can calculate M_2.

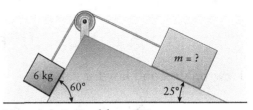

Figure 4-9 Problem 77

SOLVE

Free-body diagram of M_1:

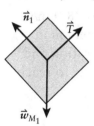

Figure 4-10 Problem 77

Newton's second law for M_1:

$$\sum F_{\text{ext},\parallel} = T - w_{M_1,\parallel} = T - M_1 g \sin(60.0°) = M_1 a_\parallel = 0$$

$$T = M_1 g \sin(60.0°)$$

Free-body diagram for M_2:

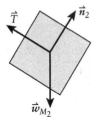

Figure 4-11 Problem 77

Newton's second law for M_2:

$$\sum F_{\text{ext},\parallel} = -T + w_{2,\parallel} = -T + M_2 g \sin(25°) = M_2 a_\parallel = 0$$

$$-M_1 g \sin(60.0°) + M_2 g \sin(25°) = 0$$

$$M_2 = \frac{M_1 \sin(60.0°)}{\sin(25°)} = \frac{(6.00 \text{ kg}) \sin(60.0°)}{\sin(25°)} = \boxed{12.3 \text{ kg}}$$

REFLECT

The angle on the left is steeper than the angle on the right, so the mass on the left should be smaller than the mass on the right.

Get Help: Interactive Exercise – Tension
Interactive Exercise – Two Strings

4.81

SET UP

A 30.0-kg dog stands on a scale, and both are placed in an elevator. The elevator begins to move and we are interested in calculating the reading on the scale for various accelerations of the elevator, which are equal to the accelerations of the scale and dog since they travel together as one object. The reading on the scale is equal in magnitude to the normal force acting on the dog. Gravity also acts on the dog. We can use Newton's second law to calculate N as a function of the acceleration. We'll define up to be positive y throughout our calculation.

SOLVE

Free-body diagram of the dog:

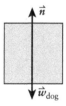

Figure 4-12 Problem 81

$$\sum F_{ext,y} = n - w_{dog} = m_{dog} a_y$$

$$n = w_{dog} + m_{dog} a_y = m_{dog} g + m_{dog} a_y = m_{dog}(g + a_y)$$

Part a)

$$n = m_{dog}(g + a_y) = (30.0 \text{ kg})\left(\left(9.80 \frac{m}{s^2}\right) - \left(3.50 \frac{m}{s^2}\right)\right) = \boxed{189 \text{ N}}$$

Part b)

$$n = m_{dog}(g + a_y) = (30.0 \text{ kg})\left(\left(9.80 \frac{m}{s^2}\right) + 0\right) = \boxed{294 \text{ N}}$$

Part c)

$$n = m_{dog}(g + a_y) = (30.0 \text{ kg})\left(\left(9.80 \frac{m}{s^2}\right) + \left(4.00 \frac{m}{s^2}\right)\right) = \boxed{414 \text{ N}}$$

REFLECT

When an elevator starts to move downward from rest, you feel lighter; when an elevator travels at a constant speed you feel "normal"; and, when an elevator moves upward from rest, you feel heavier. All of these observations from everyday life correspond to our calculation with respect to the dog and the scale.

Get Help: Picture It – Apparent Weight

4.85

SET UP

A froghopper has a mass of 12.3 mg and can flex its legs 2.0 mm in order to jump an additional 426 mm to a height of 428 mm above the ground. We are interested in the portion of the jump while the insect is still on the ground. We can assume that the acceleration of the insect is constant during this phase. In order to calculate the acceleration of the insect and the length of time during this phase of the jump we first need to calculate the speed with which the froghopper leaves the ground. While the insect is in the air, it is only under the influence of gravity, so we can use the constant acceleration equations and the height of the jump to calculate the takeoff speed. Once we have this value, we know the froghopper accelerated from rest through a distance of 2.00 mm to reach this takeoff speed. We can directly calculate the time because we are assuming the acceleration is constant. The forces acting on the insect while it is on the ground are the force of the ground on the froghopper pointing up (that is, the normal force) and the force of gravity pointing down. After defining a coordinate system where positive y points upward, we can calculate the magnitude of the normal force using Newton's second law.

SOLVE

Part a)
Takeoff speed:

$$v_y^2 = v_{0y}^2 + 2a_y(y - y_0)$$

$$v_{0y} = \sqrt{v_y^2 - 2a_y(\Delta y)} = \sqrt{0 - 2\left(-9.80\frac{m}{s^2}\right)(0.426 \text{ m})} = 2.89\frac{m}{s}$$

Acceleration necessary to attain that speed from rest:

$$v_y^2 = v_{0y}^2 + 2a_y(y - y_0)$$

$$a_y = \frac{v_y^2 - v_{0y}^2}{2(\Delta y)} = \frac{\left(2.89\frac{m}{s}\right)^2 - 0}{2(0.00200 \text{ m})} = \boxed{2.09 \times 10^3 \frac{m}{s^2}}$$

Time:

$$a_y = \frac{\Delta v_y}{\Delta t}$$

$$\Delta t = \frac{\Delta v_y}{a_y} = \frac{2.89\frac{m}{s}}{2.09 \times 10^3 \frac{m}{s^2}} = \boxed{0.00138 \text{ s} = 1.38 \text{ ms}}$$

Part b)

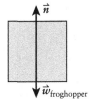

Figure 4-13 Problem 85

Part c)

$$\sum F_{ext,y} = n - w_{froghopper} = n - m_{froghopper}g = m_{froghopper}a_y$$

$$n = m_{froghopper}g + m_{froghopper}a_y = m_{froghopper}(g + a_y)$$

$$= (12.3 \times 10^{-6} \text{ kg})\left(\left(9.80 \frac{m}{s^2}\right) + \left(2.09 \times 10^3 \frac{m}{s^2}\right)\right)$$

$$= \boxed{0.0258 \text{ N} = 25.8 \text{ mN}}$$

This force is $\boxed{214 \text{ times larger}}$ than the froghopper's weight.

REFLECT

A froghopper is around 5mm long, so a height of 426 mm is about 85 times its length.

Get Help: P'Cast 4.10 – Down the Slopes

4.87

SET UP

Sue and Paul are attached by a rope while climbing a glacier with a 45.0° slope. Suddenly, Sue (m_{Sue}=66.0 kg) falls into a crevasse and falls 2.00 m in 10.0 s from rest. We'll use two different but related coordinate systems for the two people. For Sue, positive y will point upward. For Paul, the axes will be parallel and perpendicular to the inclined plane, where up the ramp and out of the ramp are positive. Tension from the rope pulling up and gravity pulling down are the only forces acting on Sue. Assuming her acceleration is constant, we can use the constant acceleration equations and Newton's second law to calculate the magnitude of the tension. Since Paul and Sue are tethered to one another, the magnitudes of their accelerations are equal. The tension in the rope and gravity are the only forces acting on Paul that have components that are parallel to the face of the glacier. We can then solve the parallel component of Newton's second law for Paul's mass.

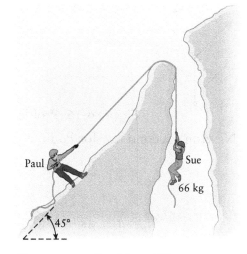

Figure 4-14 Problem 87

SOLVE

Part a)

Sue's acceleration:

$$y = y_0 + v_{0y}t + \frac{1}{2}a_y t^2 = y_0 + 0 + \frac{1}{2}a_y t^2$$

$$a_y = \frac{2(\Delta y)}{t^2} = \frac{2(-2.00 \text{ m})}{(10.0 \text{ s})^2} = -0.0400 \frac{\text{m}}{\text{s}^2}$$

Free-body diagram of Sue:

Figure 4-15 Problem 87

Newton's second law for Sue:

$$\sum F_{\text{ext},y} = T - w_{\text{Sue}} = T - m_{\text{Sue}}g = m_{\text{Sue}}a_y$$

$$T = m_{\text{Sue}}(g + a_y) = (66.0 \text{ kg})\left(\left(9.80\frac{\text{m}}{\text{s}^2}\right) + \left(-0.0400\frac{\text{m}}{\text{s}^2}\right)\right) = \boxed{644 \text{ N}}$$

Part b)

Free-body diagram of Paul:

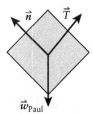

Figure 4-16 Problem 87

Newton's second law for Paul:

$$\sum F_{\text{ext},\parallel} = T - w_{\text{Paul}} = T - m_{\text{Paul}}g\sin(45.0°) = m_{\text{Paul}}a_\parallel$$

$$m_{\text{Paul}} = \frac{T}{a_\parallel + g\sin(45.0°)} = \frac{(644 \text{ N})}{\left(0.0400\frac{\text{m}}{\text{s}^2}\right) + \left(9.80\frac{\text{m}}{\text{s}^2}\right)\sin(45.0°)} = \boxed{92.4 \text{ kg}}$$

REFLECT

The tension in the rope (644 N) is less than Sue's weight (647 N), so there is a net force on Sue acting downward, which means she'll accelerate in that direction. Paul's mass of 92 kg is around 200 lb, which is a reasonable weight for a human being.

Get Help: Interactive Exercise – Tension
Interactive Exercise – Two Strings
P'Cast 4.10 – Down the Slopes

4.89

SET UP

A man sits in a bosun's chair (see figure). Since we are given the combined mass of the man, chair, and bucket ($m = 95.0$ kg), we will treat them together as one object. There are two tension forces upward on this object (one from each end of the rope) and the force of gravity acting downward. Using a coordinate system where positive y points upward, we can solve Newton's second law for the tension in terms of the acceleration. Recall that an object traveling at a constant velocity has an acceleration of zero.

Figure 4-17 Problem 89

SOLVE

Free-body diagram of the man + chair + bucket:

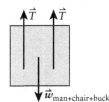

Figure 4-18 Problem 89

Newton's second law for the man + chair + bucket:

$$\sum F_{ext,y} = 2T - w_{man+chair+bucket} = 2T - m_{man+chair+bucket}g = m_{man+chair+bucket}a_y$$

$$T = \frac{m_{man+chair+bucket}(g + a_y)}{2}$$

Part a) At constant speed:

$$T = \frac{(95.0 \text{ kg})\left(\left(9.80\frac{m}{s^2}\right) + 0\right)}{2} = \boxed{466 \text{ N}}$$

Part b) At a constant upward acceleration of 1.50 m/s²:

$$T = \frac{(95.0 \text{ kg})\left(\left(9.80\frac{m}{s^2}\right) + \left(1.50\frac{m}{s^2}\right)\right)}{2} = \boxed{537 \text{ N}}$$

REFLECT

In order to remain stationary or travel at a constant speed, the man needs to pull with a force equal to the weight. The man should pull harder if he wants to accelerate upward, which makes sense.

4.91

SET UP

Blocks A and B are connected by a string with a tension T_1. Block C, which is attached to block A by a string with tension T_2, is hanging off of the table. Block A has a mass $m_A = 2.00$ kg, block B has a mass of $m_B = 1.00$ kg, and block C has a weight of 10.0 N (that is, $m_C = 1.02$ kg). We are told that blocks A and B accelerate to the right, which

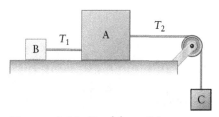

Figure 4-19 Problem 91

we'll call positive x. This means block C is falling (toward negative y). All three blocks have the same acceleration because they are tethered together. Using Newton's second law for each of the blocks, we can solve for the acceleration in terms of the known masses and weight. Once we have the acceleration, we can calculate the magnitudes of the tensions.

SOLVE

Free-body diagram of block A:

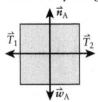

Figure 4-20 Problem 91

Free-body diagram of block B:

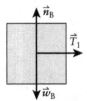

Figure 4-21 Problem 91

Free-body diagram of block C:

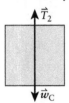

Figure 4-22 Problem 91

Newton's second law:

$$\sum F_{\text{ext},Ax} = -T_1 + T_2 = m_A a$$

$$\sum F_{\text{ext},Bx} = T_1 = m_B a$$

$$\sum F_{\text{ext},Cy} = T_2 - w_C = m_C(-a)$$

Solving for the acceleration:

$$T_2 = w_C - m_C a = (m_A + m_B)a$$

$$a = \frac{w_C}{m_A + m_B + m_C}$$

Solving for the tensions:

$$T_1 = m_B a = m_B\left(\frac{w_C}{m_A + m_B + m_C}\right) = (1.00 \text{ kg})\left(\frac{10.0 \text{ N}}{(2.00 \text{ kg}) + (1.00 \text{ kg}) + (1.02 \text{ kg})}\right) = \boxed{2.49 \text{ N}}$$

$$T_2 = (m_A + m_B)a = (m_A + m_B)\left(\frac{w_C}{m_A + m_B + m_C}\right)$$

$$= ((2.00 \text{ kg}) + (1.00 \text{ kg}))\left(\frac{10.0 \text{ N}}{(2.00 \text{ kg}) + (1.00 \text{ kg}) + (1.02 \text{ kg})}\right) = \boxed{7.46 \text{ N}}$$

REFLECT

It makes sense that $T_2 > T_1$ since the second string has to pull *both* blocks A and B, while the first string is only pulling block B. We could have also treated all three blocks as one object with a combined mass of 4 kg. The only external force acting on this system is $w_C = 10$ N. This would have quickly given us the acceleration for the system.

Get Help: Interactive Exercise – Tension
Interactive Exercise – Two Strings

Chapter 5
Forces and Motion II: Applications

Conceptual Questions

5.3 The force of static friction between two objects can cause one object to accelerate, and thus increase in speed, when the second object begins to accelerate. For example, when a pickup truck accelerates forward, a box in the bed of the truck will also accelerate forward due to the force of static friction between the box and the truck's bed.

5.7 If the water stays in the bucket, it is because the whirling is so fast that at the top of the circle, the bucket is accelerating downward faster than the acceleration of gravity. The bottom of the bucket is needed to pull the water down! The sides of the bucket prevent the water from sloshing ahead or behind with the rest of the water that is going in the same circle. The weight is always pointing straight down (whether at the top or at the bottom), but the normal force n changes direction, pointing down at the top of the swing and up at the lowest point. Thus, $\sum F_{\text{ext on top},y} = mg + n$ while $\sum F_{\text{ext on bottom},y} = mg - n$.

Get Help: P'Cast 5.8 – A Rock on a String

5.11 His acceleration decreases because the net force on him decreases. Net force is equal to his weight minus his drag force. The faster an object is moving, the larger the drag force it experiences as a function of its speed. Because drag force increases with increasing speed, net force and acceleration decrease. As the drag force approaches his weight, the acceleration approaches zero.

Get Help: P'Cast 5.7 – Terminal Speed

5.15 If you know the car's instantaneous speed and the radius of curvature, you can use the formula $a_{\text{cent}} = \dfrac{v^2}{r}$. A more direct alternative is to use an accelerometer—a circular glass tube with a ball in it. If the ball settles at angle α from the bottom, tan α = (centripetal acceleration)/(acceleration of gravity).

Get Help: P'Cast 5.10 – A Turn on a Level Road

5.17 Banking the curve allows the normal force to help push the car into the turn. The additional normal force also increases the maximum frictional force.

Multiple-Choice Questions

5.21 **A** (slide down at constant speed).

Free-body diagram of a mass sliding down a ramp:

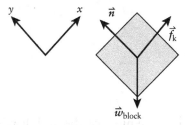

Figure 5-1 Problem 21

We take the positive x direction to point up the ramp and the positive y direction to point perpendicular to the ramp.

Newton's second law:

$$\sum F_{ext,y} = n - m_{block}g\cos(\theta) = m_{block}a_{ext,y} = 0$$

$$n = m_{block}g\cos(\theta)$$

$$\sum F_{ext,x} = \mu_k n - m_{block}g\sin(\theta) = m_{block}a_{ext,x}$$

$$\mu_k(m_{block}g\cos(\theta)) - m_{block}g\sin(\theta) = m_{block}a_{ext,x}$$

$$a_{ext,x} = \mu_k g\cos(\theta) - g\sin(\theta)$$

The acceleration is independent of the mass, which means the blocks will have the same motion.

5.25 **D** (the drag force is larger than the skydiver's weight). Once the skydiver opens his parachute, he will start to slow down because there is a net force acting upward. A net force upward means that the magnitude of the drag force is larger than the magnitude of his weight.

Get Help: P'Cast 5.7 – Terminal Speed

5.27 **B** ($n > mg$). There is a net force acting upward because there is an acceleration in that direction when you are at the bottom of a rotating Ferris wheel.

Estimation/Numerical Analysis

5.31 Assume the initial speed of the runner to be 7 m/s and the slide covers 3 m. The acceleration would be:

$$v_x^2 = v_{0x}^2 + 2a_x(x - x_0)$$

$$a_x = \frac{v_x^2 - v_{0x}^2}{2\Delta x} = \frac{0 - \left(7\frac{m}{s}\right)^2}{2(3\text{ m})} = -8\frac{m}{s^2}$$

48 Chapter 5 Forces and Motion II: Applications

$$\sum F_{ext,x} = \mu_k m_{runner} g = m_{runner} a_x$$

$$\mu_k = \frac{a_x}{g} = \frac{\left(-8\frac{m}{s^2}\right)}{\left(-9.80\frac{m}{s^2}\right)} = \boxed{0.8}$$

5.35 It takes about 2 s to fall 2 m at a 30° angle from vertical (the length of the slide); with these data, the length of the slide is 4 m and your acceleration is 2 m/s². The coefficient of kinetic friction is approximately 0.3.

Get Help: P'Cast 5.6 – A Sled Ride

Problems

5.39

SET UP

A 7.60-kg object rests on a level floor with a coefficient of static friction of 0.550. The minimum horizontal applied force that will cause the object to start sliding is equal to the maximum possible magnitude of the static frictional force. This will have a magnitude equal to the coefficient of static friction multiplied by the normal force acting on the object. Because the object is stationary in the vertical direction, the normal force is equal in magnitude to the weight.

SOLVE

$$F_{applied} = f_{s,max} = \mu_s n = \mu_s m_{object} g = (0.550)(7.60 \text{ kg})\left(9.80\frac{m}{s^2}\right) = \boxed{41.0 \text{ N}}$$

REFLECT

Usually the magnitude of static friction involves an inequality. We can use an equal sign here because we're looking for the maximum static frictional force.

5.41

SET UP

A book is pushed across a horizontal table with a force equal to one-half the book's weight. The book is traveling at a constant speed, which means its acceleration in this direction is equal to zero. The only forces acting in the horizontal direction are the applied force and kinetic friction, and they act in opposite directions. The magnitude of the kinetic frictional force is constant and equal to the product of the coefficient of kinetic friction and the normal force. Assuming that the object is not accelerating in the vertical direction, the magnitude of the normal force is equal to the object's weight. Setting the forces equal we can solve for the coefficient of kinetic friction.

SOLVE

$$\sum F_{ext,x} = F_{applied} - f_k = m_{book} a_x = 0$$

$$F_{applied} = f_k = \mu_k n = \mu_k w_{book}$$

$$\mu_k = \frac{F_{applied}}{w_{book}} = \frac{\left(\frac{w_{book}}{2}\right)}{w_{book}} = \boxed{0.5}$$

REFLECT

This is a reasonable value for the coefficient of kinetic friction for a book sliding across a table.

Get Help: Interactive Exercise – Wooden Block

5.47

SET UP

A 12.0-kg block and a 5.00-kg block are connected by a taut string with tension T. The coefficient of static friction between the 12.0-kg block and the floor is $\mu_{s,M2} = 0.443$ and the coefficient of static friction between the 5.00-kg block and the floor is $\mu_{s,M1} = 0.573$. An external force F pulling to the right is applied to the 5.00-kg block. We need to determine, using Newton's second law, the minimum F necessary to overcome static friction and accelerate the blocks toward positive x. Since the mass is positive and the x component of the acceleration will be positive when the blocks move, the net force acting on each block also needs to be positive (that is, greater than zero).

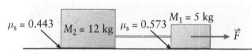

Figure 5-2 Problem 47

SOLVE

Free-body diagram of M_2:

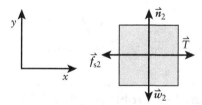

Figure 5-3 Problem 47

Free-body diagram of M_1:

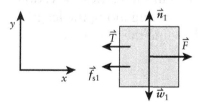

Figure 5-4 Problem 47

Newton's second law, y components:

$$\sum F_{\text{ext on 2},y} = n_2 - w_2 = n_2 - M_2 g = M_2 a_y = 0$$

$$n_2 = M_2 g$$

$$\sum F_{\text{ext on 1},y} = n_1 - w_1 = n_1 - M_1 g = M_1 a_y = 0$$

$$n_1 = M_1 g$$

Newton's second law, x components:

$$\sum F_{\text{ext on 2},x} = T - f_{s2} = T - \mu_{s,2} n_2 = T - \mu_{s,2} M_2 g = M_2 a_x$$

$$\sum F_{\text{ext on 1},x} = F - T - f_{s1} = F - T - \mu_{s,1} n_1 = F - T - \mu_{s,1} M_1 g = M_1 a_x$$

Adding the x-component equations together:

$$F - g(\mu_{s,2} M_2 + \mu_{s,1} M_1) = (M_1 + M_2) a_x$$

Setting up and solving the inequality:

$$F - g(\mu_{s,2} M_2 + \mu_{s,1} M_1) = (M_1 + M_2) a_x \geq 0$$

$$F \geq g(\mu_{s,2} M_2 + \mu_{s,1} M_1) = \left(9.80 \frac{\text{m}}{\text{s}^2}\right)((0.443)(12.0 \text{ kg}) + (0.573)(5.00 \text{ kg})) = \boxed{80.2 \text{ N}}$$

REFLECT

The maximum value of static friction is when $f_{s,\max} = \mu_s n$. When the magnitude of the applied force is larger than this value, the system will start to accelerate.

Get Help: P'Cast 5.6 – Pinned against a Wall

5.49

SET UP

Two blocks are connected by a string. Block 1 is hanging over the edge of a ramp. Block 2 ($m_2 = 8.00$ kg) is resting on a 28.0° slope. The coefficient of kinetic friction between block 2 and the ramp is 0.22. We are told block 2 is sliding down the ramp at a constant speed, which means its acceleration is zero in all directions. Because the two blocks

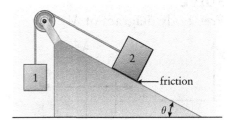

Figure 5-5 Problems 49 and 50

are connected, block 1 is also traveling at a constant speed. We can use Newton's second law to calculate the mass of block 1. For block 1, we will define the x-axis in the vertical direction where upwards is positive. For block 2, we will define the x-axis as pointing along the length of the ramp, and down the ramp will be positive. Similarly, we will define the y-axis as pointing perpendicular to the ramp.

SOLVE

Free-body diagram of block 1:

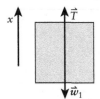

Figure 5-6 Problem 49

Free-body diagram of block 2:

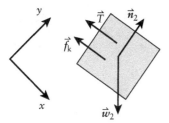

Figure 5-7 Problem 49

Newton's second law:

$$\sum F_{\text{ext on 1},x} = T - w_1 = T - m_1 g = m_1 a_x = 0$$

$$T = m_1 g$$

$$\sum F_{\text{ext on 2},y} = n_2 - w_{2y} = n_2 - m_2 g \cos(28.0°) = m_2 a_y = 0$$

$$n_2 = m_2 g \cos(28.0°)$$

$$\sum F_{\text{ext on 2},x} = w_{2x} - T - f_k = m_2 g \sin(28.0°) - T - \mu_k n_2$$

$$= m_2 g \sin(28.0°) - T - \mu_k (m_2 g \cos(28.0°))$$

$$= m_2 g \sin(28.0°) - m_1 g - \mu_k (m_2 g \cos(28.0°)) = 0$$

$$m_1 = m_2 (\sin(28.0°) - \mu_k \cos(28.0°)) = (8.00 \text{ kg})(\sin(28.0°) - (0.220)\cos(28.0°))$$

$$= \boxed{2.20 \text{ kg}}$$

REFLECT

We are told block 2 is sliding down the ramp, so we should expect the mass of block 1 to be less than 8 kg; our answer makes sense.

Get Help: Interactive Exercise – Wooden Block

5.53

SET UP

A horizontal force $F_{\text{applied}} = 10.0$ N is applied to a stationary 2.00-kg block. The coefficient of static friction between the block and the floor is 0.750. In order to determine the motion of the block, we first need to determine which friction—static or kinetic—needs to be considered. Newton's second law in the vertical direction will give us the magnitude of the normal force, which is used in calculating the magnitudes of the frictional forces. We should first calculate the magnitude of the static frictional force. If it is larger than 10.0 N, the block won't move; if it's less than 10.0 N, then we calculate the magnitude of kinetic friction and use that to determine the acceleration in the horizontal direction.

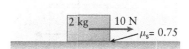

Figure 5-8 Problem 53

52 Chapter 5 Forces and Motion II: Applications

SOLVE
Free-body diagram of block:

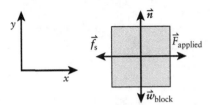

Figure 5-9 Problem 53

Newton's second law:

$$\sum F_{ext,y} = n - w_{block} = n - m_{block}g = m_{block}a_y = 0$$

$$n = m_{block}g$$

Magnitude of static friction:

$$f_s \leq \mu_s n = \mu_s m_{block} g = (0.750)(2.00 \text{ kg})\left(9.80 \frac{\text{m}}{\text{s}^2}\right) = 14.7 \text{ N}$$

The applied force of 10.0 N is not enough to overcome static friction, which means the block will remain stationary.

REFLECT
The information regarding kinetic friction is irrelevant for this problem.

Get Help: P'Cast 5.6 – Pinned against a Wall

5.59

SET UP
A 1500-kg truck is driving around an unbanked curve at a speed of 20.0 m/s. The maximum frictional force between the road and the tires is 8000 N. We can use Newton's second law to relate the maximum frictional force to the radius of the curvature by realizing the car is undergoing uniform circular motion and that static friction is the force responsible for the car traveling around the circle.

SOLVE
Free-body diagram of the truck:

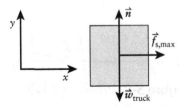

Figure 5-10 Problem 59

Newton's second law:

$$\sum F_{ext,y} = n - w_{truck} = n - m_{truck}g = m_{truck}a_y = 0$$

$$n = m_{truck}g$$

$$\sum F_{ext,x} = f_{s,max} = m_{truck} a_{cent} = m_{truck}\left(\frac{v^2}{r}\right)$$

$$r = \frac{m_{truck} v^2}{f_{s,max}} = \frac{(1500 \text{ kg})\left(20.0\frac{\text{m}}{\text{s}}\right)^2}{8000 \text{ N}} = \boxed{8 \times 10^1 \text{ m}}$$

REFLECT

Although it may seem large, this is a reasonable radius for a curve on a highway. The curves on a highway are usually gentle because cars are traveling at high speeds.

Get Help: P'Cast 5.10 – A Turn on a Level Road

5.61

SET UP

A centrifuge of radius $r = 0.100$ m spins a 1.00-g sample at 1200 rev/min. The net force keeping the sample on its circular path is equal to the mass of the sample multiplied by the centripetal acceleration of the sample.

SOLVE

$$F_{\text{centrifuge on sample}} = m_{sample} a_{cent} = \frac{m_{sample} v^2}{r}$$

$$= \frac{(1.00 \times 10^{-3} \text{ kg})\left(\frac{1200 \text{ rev}}{\text{min}} \times \frac{1 \text{ min}}{60 \text{ s}} \times \frac{2\pi(0.100 \text{ m})}{1 \text{ rev}}\right)^2}{0.100 \text{ m}} = \boxed{1.6 \text{ N}}$$

REFLECT

This force is about 100 times larger than the weight of the sample.

5.65

SET UP

A 0.750-kg tetherball is attached to a rope of length $L = 1.25$ m and is spun around a pole. The rope makes an angle of 35.0° with the vertical. Although the ball is traveling in a circle in the horizontal plane, the ball is at rest with respect to the vertical direction. We can use Newton's second law to calculate the tension in the rope. After we solve for tension, we can determine the horizontal component of the tension to calculate the speed of the ball as it travels in a circle.

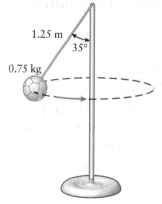

Figure 5-11 Problem 65

54 Chapter 5 Forces and Motion II: Applications

SOLVE
Free-body diagram of the ball:

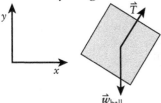

Figure 5-12 Problem 65

Newton's second law:

$$\sum F_{ext,y} = T_y - w_{ball} = T\cos(35.0°) - m_{ball}g = m_{ball}a_y = 0$$

$$T = \frac{m_{ball}g}{\cos(35.0°)} = \frac{(0.750 \text{ kg})\left(9.80\frac{\text{m}}{\text{s}^2}\right)}{\cos(35.0°)} = \boxed{8.97 \text{ N}}$$

$$\sum F_{ext,x} = T\sin(35.0°) = m_{ball}a_{cent} = \frac{m_{ball}v^2}{r} = \frac{m_{ball}v^2}{L\sin(35.0°)}$$

$$v = \sqrt{\frac{TL}{m_{ball}}\sin(35.0°)} = \sqrt{\frac{(8.973 \text{ N})(1.25 \text{ m})}{0.750 \text{ kg}}\sin(35.0°)} = 2.22\frac{\text{m}}{\text{s}}$$

REFLECT
Only the horizontal component of the tension contributes to the circular motion of the ball around the pole.

Get Help: P'Cast 5.8 – A Rock on a String

General Problems

5.71

SET UP
The coefficient of static friction between a rubber tire and dry pavement is around 0.800. We can use Newton's second law in order to calculate the acceleration of the car and then the time required for the car to accelerate from rest to 26.8 m/s. We are told that the engine supplies power to only two of the four wheels, which means we need to divide the maximum force due to static friction by two. Friction opposes relative *slipping* motion between two objects but can still cause the car to accelerate forward.

SOLVE

Part a)
Free-body diagram of the car:

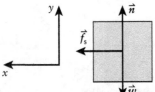

Figure 5-13 Problem 71

Newton's second law:

$$\sum F_{ext,y} = n - w_{car} = n - m_{car}g = m_{car}a_y = 0$$

$$n = m_{car}g$$

$$\sum F_{ext,x} = f_s = \mu_s\left(\frac{n}{2}\right) = \mu_s\left(\frac{m_{car}g}{2}\right) = m_{car}a_x$$

$$a_x = \frac{\mu_s g}{2} = \frac{(0.800)\left(9.80\frac{m}{s^2}\right)}{2} = 3.92\frac{m}{s^2}$$

Calculating the time:

$$a_x = \frac{\Delta v_x}{\Delta t}$$

$$\Delta t = \frac{\Delta v_x}{a_x} = \frac{\left(26.8\frac{m}{s}\right)}{\left(3.92\frac{m}{s^2}\right)} = \boxed{6.84 \text{ s}}$$

Part b) Static friction opposes the relative slipping motion between two surfaces. The bottom of the tire tends to slip backward relative to the wet pavement, so the frictional force opposes the backward motion and, hence, is forward.

REFLECT

On dry pavement, the car's tires do not slip relative to the surface of the road. An example of when the tires *do* slip would be on an icy road or on rain-soaked pavement.

Get Help: P'Cast 5.6 – A Sled Ride

5.75

SET UP

A 2.50-kg package slides down a 20.0° incline with an initial speed of 2.00 m/s. The package slides a total distance of 12.0 m. We can calculate the constant acceleration of the package assuming it comes to rest at the bottom of the ramp. Using Newton's second law, we can relate this acceleration to the net force on the package and then calculate the coefficient of kinetic friction between the package and the ramp. We can define the *x*-axis along the length of the incline, and the positive direction is down the incline. The *y*-axis points perpendicular to the incline.

SOLVE

Finding the acceleration:

$$v_x^2 = v_{0x}^2 + 2a_x(x - x_0)$$

$$a_x = \frac{v_x^2 - v_{0x}^2}{2(\Delta x)} = \frac{0 - \left(2.00\frac{m}{s}\right)^2}{2(12.0 \text{ m})} = -0.167\frac{m}{s^2}$$

56 Chapter 5 Forces and Motion II: Applications

Free-body diagram of the package:

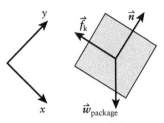

Figure 5-14 Problem 75

Newton's second law:

$$\sum F_{ext,y} = n - w_{package,y} = n - m_{package}g\cos(20.0°) = m_{package}a_y = 0$$

$$n = m_{package}g\cos(20.0°)$$

$$\sum F_{ext,x} = w_{package,x} - f_k = m_{package}g\sin(20.0°) - \mu_k n$$

$$= m_{package}g\sin(20.0°) - \mu_k m_{package}g\cos(20.0°) = ma_x$$

$$\mu_k = \frac{g\sin(20.0°) - a_x}{g\cos(20.0°)} = \frac{\left(9.80\frac{m}{s^2}\right)\sin(20.0°) - \left(-0.167\frac{m}{s^2}\right)}{\left(9.80\frac{m}{s^2}\right)\cos(20.0°)} = \boxed{0.382}$$

REFLECT

We can look at the limiting cases in order to double-check our algebraic solution. As the angle of the ramp approaches 0, the magnitude of kinetic friction should be smaller if the block comes to rest at the end of the ramp. As the angle of the ramp approaches 90°, then the coefficient of kinetic friction will need to be extremely large in order to attain the same acceleration.

5.81

SET UP

A mass M is attached to a string of length L and is rotated. The angle the string makes with the vertical is θ. The speed of the mass is N revolutions per second. In one revolution the mass sweeps out a circle of radius $r = L\sin(\theta)$. We can use Newton's second law to solve for the observed value of θ. The only forces acting on the mass are tension from the string and gravity. The mass is stationary in the y direction and undergoing centripetal motion in the x direction. Keep in mind that we are only allowed to have M, L, N, and physical constants in our final answer.

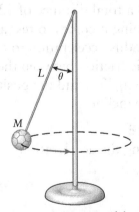

Figure 5-15 Problem 81

SOLVE
Free-body diagram of M:

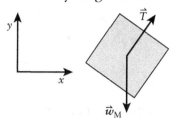

Figure 5-16 Problem 81

Newton's second law:

$$\sum F_{ext,x} = T_x = T\sin(\theta) = M\left(\frac{v^2}{r}\right) = M\left(\frac{(N(2\pi r))^2}{r}\right) = M(4\pi^2 N^2 r) = 4\pi^2 N^2 M(L\sin(\theta))$$

$$T = 4\pi^2 N^2 ML$$

$$\sum F_{ext,y} = T_y - w_M = T\cos(\theta) - Mg = Ma_y = 0$$

$$T\cos(\theta) = 4\pi^2 N^2 ML\cos(\theta) = Mg$$

$$\cos(\theta) = \frac{g}{4\pi^2 N^2 L}$$

$$\boxed{\theta = \arccos\left(\frac{g}{4\pi^2 N^2 L}\right)}$$

REFLECT
If we observe that the mass makes a smaller angle with the vertical, we would assume the rotation speed has increased. It also makes sense that a smaller angle could result from a longer string (that is, increased L).

Get Help: P'Cast 5.8 – A Rock on a String

Chapter 6
Work and Energy

Conceptual Questions

6.3 Stepping on the log would require raising the body, which would require doing work on it. Most of the work would not be recovered when coming down off the log.

6.7 If the surface is moving, yes. Otherwise, the distance over which the force is applied is zero, so the work will be zero.

6.11 The snowboarder walks to the top of the mountain, which requires converting stored chemical energy into heat and kinetic energy, which in turn is converted into gravitational potential energy and more heat (if the snowboarder rides a lift, the source will also include wherever the lift gets its energy). Then, riding down, the gravitational potential energy and some chemical energy will be converted into kinetic energy, which is in turn dissipated into heat.

6.15 (a) When it reaches the bottom. (b) When it reaches the bottom (but it's been close to that speed for a while).

Multiple-Choice Questions

6.21 C (decreases). As θ increases to 90°, $\cos(\theta)$ approaches zero.

Get Help: P'Cast 6.1 – Lifting a Book
P'Cast 6.2 – Work Done by Actin

6.25 C (They have the same kinetic energy.) Each twin starts with the same gravitational potential energy and falls the same distance.

Estimation/Numerical Analysis

6.29 A standard estimate of the power of hard jogging is 7–8 kcal/kg/h. For a runner who is 80 kg and who runs for 25 min, the energy is roughly 1 MJ.

Get Help: Picture It – Mechanical Energy
Interactive Exercise – Bobsled

6.33 Pumas have been seen to leap 5.4 m. If the puma's average mass is 62 kg, the maximum kinetic energy is about 3000 J.

Problems

6.37

SET UP

A weightlifter lifts 446 kg a distance of 2.0 m. If the mass is moving at constant velocity, then the magnitude of the force of the weightlifter on the mass is equal to the weight of the mass. Both of these forces are constant. The force of the weightlifter on the mass is parallel to the displacement of the mass.

SOLVE

$$W_{\text{weightlifter}} = F_{\text{weightlifter}} d \cos(0°) = mgd = (446 \text{ kg})\left(9.80 \frac{\text{m}}{\text{s}^2}\right)(2.0 \text{ m}) = \boxed{8.7 \times 10^3 \text{ J}}$$

REFLECT

We are only allowed two significant figures in our final answer since d has two significant figures. The force and the displacement are in the same direction, so we expect W to be positive.

Get Help: P'Cast 6.1 – Lifting a Book
P'Cast 6.2 – Work Done by Actin

6.41

SET UP

A 150-kg crate is moving down a 40.0° incline. The coefficient of kinetic friction between the crate and the ramp is 0.54. A museum curator is pushing against the crate with enough force that the crate does not accelerate. We can use Newton's second law to determine the magnitude of the curator's force and the magnitude of kinetic friction. Once we know the magnitudes and directions of all of the forces acting on the crate, we need to determine the angle between each force and the displacement of the crate before calculating the work done by each force.

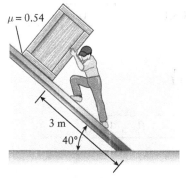

Figure 6-1 Problem 41

SOLVE
Free-body diagram of the crate:

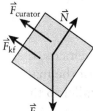

Figure 6-2 Problem 41

Part a)
$$w_{\text{gravity}} = wd\cos(\theta) = wd\cos(50.0°) = mgd\cos(50.0°)$$
$$= (150 \text{ kg})\left(9.80\frac{\text{m}}{\text{s}^2}\right)(3.0 \text{ m})\cos(50.0°) = \boxed{2.8 \times 10^3 \text{ J}}$$

Part b)
Newton's second law:
$$\sum F_{\text{ext},\perp} = n - w_{\perp} = n - mg\cos(40.0°) = ma_{\perp} = 0$$
$$n = mg\cos(40.0°)$$
$$\sum F_{\text{ext},\parallel} = w_{\parallel} - F_{\text{curator}} - f_k = mg\sin(40°) - F_{\text{curator}} - \mu_k n$$
$$= mg\sin(40.0°) - F_{\text{curator}} - \mu_k(mg\cos(40.0°)) = ma_{\parallel} = 0$$
$$F_{\text{curator}} = mg(\sin(40.0°) - \mu_k\cos(40.0°))$$

Calculating the work:
$$W_{\text{curator}} = F_{\text{curator}} d\cos(180°) = -mg(\sin(40.0°) - \mu_k\cos(40.0°))d$$
$$= -(150 \text{ kg})\left(9.80\frac{\text{m}}{\text{s}^2}\right)(\sin(40.0°) - (0.54)\cos(40.0°))(3.0 \text{ m}) = \boxed{-1.0 \times 10^3 \text{ J}}$$

Part c)
$$W_{\text{fric}} = f_k d\cos(180°) = -(\mu_k n)d = -\mu_k(mg\cos(40.0°))d$$
$$= -(0.54)(150 \text{ kg})\left(9.80\frac{\text{m}}{\text{s}^2}\right)\cos(40.0°)(3.0 \text{ m}) = \boxed{-1.8 \times 10^3 \text{ J}}$$

Part d)
$$W_n = nd\cos(90°) = \boxed{0}$$

REFLECT
Remember that the angle used when calculating work is the angle between the force vector and the displacement vector, not the angle of the inclined plane.

Get Help: P'Cast 6.1 – Lifting a Book
P'Cast 6.2 – Work Done by Actin

6.43

SET UP

A 0.25-g bumblebee is moving at a speed of 10 m/s. We can calculate its kinetic energy directly from these quantities.

SOLVE

$$K = \frac{1}{2}mv^2 = \frac{1}{2}(2.5 \times 10^{-4} \text{ kg})\left(10\frac{\text{m}}{\text{s}}\right)^2 = \boxed{1 \times 10^{-2} \text{ J}}$$

REFLECT

The speed has one significant figure, so our answer has one significant figure.

6.45

SET UP

A 2.00-kg block, which is initially at rest, is pushed by a force of 40.0 N for 22.0 m over a frictionless surface. This is the only force that does work on the block because the normal force and the force due to gravity are perpendicular to the displacement of the block. The net work on the block is equal to the change in its kinetic energy; this allows us to calculate the block's final speed.

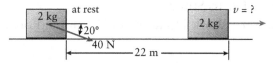

Figure 6-3 Problem 45

SOLVE

$$W_{push} = F_{push} d \cos(20.0°) = (40.0 \text{ N})(22.0 \text{ N})\cos(20.0°) = 267 \text{ J}$$

$$W_{net} = \Delta K = K_f - K_i = \frac{1}{2}mv_f^2 - \frac{1}{2}mv_i^2 = \frac{1}{2}m(v_f^2 - v_i^2) = \frac{1}{2}mv_f^2 - 0$$

$$v_f = \sqrt{\frac{2W_{net}}{m}} = \sqrt{\frac{2(827 \text{ J})}{2 \text{ kg}}} = \boxed{28.8 \frac{\text{m}}{\text{s}}}$$

REFLECT

The final speed would be larger if the force were applied at an angle closer to 0.

6.51

SET UP

A 0.145-kg baseball is initially traveling at a speed of 44.0 m/s when it is caught by a catcher's glove. The glove slows the ball to rest over a distance of 0.125 m. The net work on the ball is equal to the work done by the glove on the ball. The force of the glove on the ball points in the opposite direction to the ball's displacement. We can use the work–kinetic energy theorem to calculate the magnitude of the average force of the glove during the catch.

SOLVE

$$\Delta K = \frac{1}{2}mv_f^2 - \frac{1}{2}mv_i^2 = 0 - \frac{1}{2}mv_i^2$$

$$W_{net} = W_{glove\ on\ ball} = F_{glove\ on\ ball}d\cos(180°) = -F_{glove\ on\ ball}d$$

Using the work–kinetic energy theorem:

$$-\frac{1}{2}mv_i^2 = -F_{glove\ on\ ball}d$$

$$F_{glove\ on\ ball} = \frac{mv_i^2}{2d} = \frac{(0.145\ \text{kg})\left(44.0\frac{\text{m}}{\text{s}}\right)^2}{2(0.125\ \text{m})} = \boxed{1.12 \times 10^3\ \text{N}}$$

The force of the glove on the hand will be $\boxed{\text{slightly less}}$ than the force of the glove on the ball because the glove will compress somewhat and absorb energy.

REFLECT

The force of the glove on the ball may not necessarily be (and probably is not) constant. The average force, however, is constant.

Get Help: Interactive Exercise – Truck
P'Cast 6.5 – How Far to Stop?
P'Cast 6.7 – Find the Work Done by an Unknown Force

6.53

SET UP

An object is attached to a spring with a spring constant $k = 450\frac{\text{N}}{\text{m}}$. The object starts at a position $x_1 = 0.12$ m from equilibrium and is pulled to a position $x_2 = 0.18$ m from equilibrium. The work done by the object on the spring in stretching it is given by $W = \frac{1}{2}kx_2^2 - \frac{1}{2}kx_1^2$. The work done by the spring on the object is equal in magnitude to this but opposite in sign.

SOLVE

$$W_{spring} = -\left(\frac{1}{2}kx_2^2 - \frac{1}{2}kx_1^2\right) = -\frac{1}{2}k(x_2^2 - x_1^2)$$

$$= -\frac{1}{2}\left(450\frac{\text{N}}{\text{m}}\right)((0.18\ \text{m})^2 - (0.12\ \text{m})^2) = \boxed{-4.1\ \text{J}}$$

REFLECT

The magnitude of the work done by the spring increases nonlinearly as the object is pulled farther and farther from equilibrium. For example, the magnitude of the work done in moving the object from $x = 12$ cm to $x = 18$ cm will be less than the magnitude of the work done in moving the object from $x = 18$ cm to $x = 24$ cm, even though the displacement is +6 cm in both cases.

Get Help: Picture It – Work by a Spring

6.57

SET UP

A 40.0-kg boy glides down a hill on a skateboard. He ends up a vertical distance of 4.35 m below his initial location. The change in his gravitational potential energy is equal to this change in height multiplied by his weight.

SOLVE

$$\Delta U_{gravity} = mg\Delta y = (40.0 \text{ kg})\left(9.80\frac{\text{m}}{\text{s}^2}\right)(-4.35 \text{ m}) = \boxed{-1.71 \times 10^3 \text{ J} = -1.71 \text{ kJ}}$$

REFLECT

The boy's initial gravitational potential energy is converted into kinetic energy as he glides down the hill.

6.61

SET UP

A 0.0335-kg coin is initially at the ground level of a building and then carried up to a height of 630 m. The change in the coin's gravitational potential energy is equal to its weight multiplied by the difference in its vertical position.

SOLVE

$$\Delta U_{gravity} = mg\Delta y = (0.0335 \text{ kg})\left(9.80\frac{\text{m}}{\text{s}^2}\right)((630 \text{ m}) - (0 \text{ m})) = \boxed{2.1 \times 10^2 \text{ J}}$$

REFLECT

Technically, the answer will be a little less than this since g decreases with height.

6.65

SET UP

A water balloon is thrown straight down with an initial speed of 12.0 m/s from a height of 5.00 m above the ground. We can use conservation of energy to calculate the speed of the ball right before it hits the ground. We'll define the ground to have a potential energy of zero.

SOLVE

$$U_i + K_i = U_f + K_f$$

$$mgy_i + \frac{1}{2}mv_i^2 = 0 + \frac{1}{2}mv_f^2$$

$$v_f = \sqrt{2gy_i + v_i^2} = \sqrt{2\left(9.80\frac{\text{m}}{\text{s}^2}\right)(5.00 \text{ m}) + \left(12.0\frac{\text{m}}{\text{s}}\right)^2} = \boxed{15.6\frac{\text{m}}{\text{s}}}$$

REFLECT

This is the same answer we would get if we used the constant acceleration equation $v_y^2 - (v_{0y})^2 = 2a_y(\Delta y)$.

Get Help: Picture It – Mechanical Energy
Interactive Exercise – Bobsled

6.69

SET UP

A pendulum is made of a ball of mass m and a string of length $L = 1.25$ m. The ball is lifted to a height such that the string makes an angle of 30.0° with respect to the vertical. All of the ball's initial gravitational potential energy is converted to kinetic energy at the bottom of its swing. Applying conservation of mechanical energy, we can determine the speed of the ball at the bottom of its swing. We will measure the gravitational potential energy relative to the lowest part of the pendulum's swing. We can use trigonometry to relate the initial height y_i to the length L.

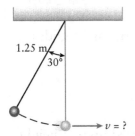

Figure 6-4 Problem 69

SOLVE

Trigonometry:

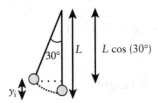

Figure 6-5 Problem 69

$$y_i = L - L\cos(30.0°) = L(1 - \cos(30.0°))$$

Conservation of energy:

$$U_i + K_i = U_f + K_f$$

$$mgy_i + 0 = 0 + \frac{1}{2}mv_f^2$$

$$gL(1 - \cos(30.0°)) = \frac{1}{2}v_f^2$$

$$v_f = \sqrt{2gL(1 - \cos(30.0°))} = \sqrt{2\left(9.80\frac{m}{s^2}\right)(1.25\text{ m})(1 - \cos(30.0°))} = \boxed{1.81\frac{m}{s}}$$

The mass impacts the speed only in that if the mass is too small, air resistance and the mass of the string will become significant.

REFLECT

The tension in the string is always perpendicular to the ball's motion; thus, the work done by the tension will be zero.

Get Help: Picture It – Mechanical Energy
Interactive Exercise – Bobsled

6.73

SET UP

A skier leaves the starting gate at an elevation of 4212 m with an initial speed of 4.00 m/s. The end of the ski slope is at an elevation of 4039 m. We are told that air resistance causes a 50% loss in the final kinetic energy (compared to the case with no air resistance). We can use conservation of mechanical energy to first determine the skier's final speed in the absence of air resistance (that is, the ideal case) and use this to calculate the skier's final speed taking air resistance into account (that is, the real case).

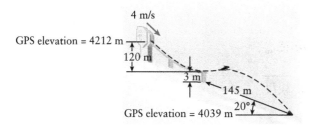

Figure 6-6 Problem 73

SOLVE

Conservation of mechanical energy:

$$U_i + K_i = U_f + K_f$$

$$\frac{1}{2}mv_i^2 + mgy_i = \frac{1}{2}mv_{f,ideal}^2 + mgy_f$$

$$v_{f,ideal}^2 = v_i^2 + 2g(y_i - y_f)$$

Considering air resistance:

$$K_{real} = \frac{1}{2}K_{ideal}$$

$$\frac{1}{2}mv_{f,real}^2 = \frac{1}{2}\left(\frac{1}{2}mv_{f,ideal}^2\right)$$

$$v_{f,real} = \sqrt{\frac{v_{f,ideal}^2}{2}} = \sqrt{\frac{v_i^2 + 2g(y_i - y_f)}{2}}$$

$$= \sqrt{\frac{\left(4.00\frac{m}{s}\right)^2 + 2\left(9.80\frac{m}{s^2}\right)((4212\ m) - (4039\ m))}{2}} = \boxed{41.3\frac{m}{s}}$$

REFLECT

The specifics regarding the middle of the skier's trip are not important since the surfaces are frictionless and we are told the change in energy caused by air resistance.

Get Help: Picture It – Mechanical Energy
Interactive Exercise – Bobsled

6.77

SET UP

The coefficient of restitution e of a ball is defined as the speed of the ball after impact divided by the speed of the ball just before impact. We can use conservation of mechanical energy to determine these speeds in terms of the heights h and H. The kinetic energy of the ball at the heights h and H is equal to zero since the ball is momentarily at rest.

SOLVE

Part a)
Before:

$$mgH = \frac{1}{2}mv_{before}^2$$

$$v_{before} = \sqrt{2gH}$$

After:

$$\frac{1}{2}mv_{after}^2 = mgh$$

$$v_{after} = \sqrt{2gh}$$

Finding the coefficient of restitution:

$$e = \frac{v_{after}}{v_{before}} = \frac{\sqrt{2gh}}{\sqrt{2gH}} = \boxed{\sqrt{\frac{h}{H}}}$$

Part b)

$$e = \sqrt{\frac{h}{H}} = \sqrt{\frac{60 \text{ cm}}{80 \text{ cm}}} = \boxed{\sqrt{\frac{3}{4}} = 0.87}$$

REFLECT

A coefficient of restitution of 1 refers to a completely elastic collision.

General Problems

6.79

SET UP

A 12.0-kg block is released from rest on a frictionless incline that makes an angle of 28.0° with the horizontal. The block starts at some initial height h_1. The block slides down the ramp, comes into contact with a spring ($k = 13{,}500$ N/m), and comes to rest once the spring compresses a distance of 5.50 cm. Since there is no nonconservative work done on the block, we can use conservation of mechanical energy. The block starts and ends at rest, which means the kinetic energies are zero and the initial potential energy will equal the final potential energy. Initially, the block is not in contact with the spring, so the initial potential energy is all gravitational. If we define the height of the compressed spring to have a gravitational potential energy of zero, the final potential energy is equal to the potential energy stored in the spring. Setting these two equal, we can find the initial height of the block relative to its stopping point. Trigonometry will allow us to find the distance that the block traveled down the ramp.

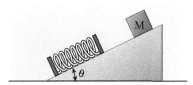

Figure 6-7 Problem 79

SOLVE

$$U_i + K_i = U_f + K_f$$

$$mgh_i + 0 = \frac{1}{2}kx^2 + 0$$

$$h_i = \frac{kx^2}{2mg}$$

$$d = \frac{h_i}{\sin(28.0°)} = \frac{kx^2}{2mg\sin(28.0°)} = \frac{\left(13{,}500\frac{\text{N}}{\text{m}}\right)(0.0550\text{ m})^2}{2(12.0\text{ kg})\left(9.80\frac{\text{m}}{\text{s}^2}\right)\sin(28.0°)} = \boxed{0.370\text{ m}}$$

REFLECT

The block starts out a distance of 31.5 cm from the front of the uncompressed spring.

6.83

SET UP

An object is released from rest on a frictionless ramp at a height of $H_1 = 12.0$ m. The bottom of this ramp merges smoothly with another frictionless ramp that makes an angle of $\theta_2 = 37.0°$ with the horizontal. Because both ramps are frictionless, the mechanical energy of the object is conserved. The block starts and ends at rest, which means the block will rise to the same height on the second ramp regardless of the angle of the ramp. We can use trigonometry to determine the distance the object travels up the second ramp. In part (b), the block starts from rest at a height of $H_1 = 12.0$ m; we can use conservation of mechanical energy to calculate its speed at a height of $H_2 = 7.00$ m.

Figure 6-8 Problem 83

SOLVE

Part a)

$$d_2 = \frac{H_2}{\sin(37.0°)} = \frac{12.0\text{ m}}{\sin(37.0°)} = \boxed{19.9\text{ m}}$$

Part b)

$$U_i + K_i = U_f + K_f$$

$$mgH_1 + 0 = mgH_2 + \frac{1}{2}mv_f^2$$

$$v_f = \sqrt{2g(H_1 - H_2)} = \sqrt{2\left(9.80\frac{m}{s^2}\right)((12.0 \text{ m}) - (7.00 \text{ m}))} = \boxed{9.90\frac{m}{s}}$$

REFLECT

The angle of the first ramp is not used because the ramp is frictionless and mechanical energy is conserved.

Get Help: Picture It – Mechanical Energy
Interactive Exercise – Bobsled

Chapter 7
Momentum, Collisions, and the Center of Mass

Conceptual Questions

7.3 Yes, if the tennis ball moves 18 times faster.
$$p_{\text{basketball}} = p_{\text{tennis ball}}$$
$$m_{\text{basketball}} v_{\text{basketball}} = m_{\text{tennis ball}} v_{\text{tennis ball}}$$
$$v_{\text{tennis ball}} = \frac{m_{\text{basketball}}}{m_{\text{tennis ball}}} v_{\text{basketball}} = \frac{(18\, m_{\text{tennis ball}})}{m_{\text{tennis ball}}} v_{\text{basketball}} = 18\, v_{\text{basketball}}$$

7.7 Pushing off the boat to step onto the pier will push the boat away. You will make less progress than on solid ground and can easily fall in the gap.
   Get Help: Picture It – Center of Mass

Multiple-Choice Questions

7.17 **D** ($2mv$). If the initial momentum of the ball points in the $-x$ direction, its initial momentum is $-mv$ and the final momentum is $+mv$. This is a total change in momentum of $2mv$.
   Get Help: P'Cast 7.2 – Conservation of Momentum: A Collision on the Ice

7.21 **B** (The magnitudes of the velocities are the same but the directions are reversed). By rearranging the energy and momentum conservation equations for two objects undergoing a one-dimensional, elastic collision, we can write the final speeds of each object:
$$\vec{P}_f = \vec{P}_i$$
$$m_A v_{Afx} + m_B v_{Bfx} = m_A v_{Aix} + m_B v_{Bix}$$
$$K_f = K_i$$
$$\frac{1}{2} m_A v_{Afx}^2 + \frac{1}{2} m_B v_{Bfx}^2 = \frac{1}{2} m_A v_{Aiy}^2 + \frac{1}{2} m_B v_{Biy}^2$$
$$v_{Afx} = \left(\frac{m_A - m_B}{m_A + m_B}\right) v_{Aix} + \left(\frac{2 m_B}{m_A + m_B}\right) v_{Bix}$$
$$v_{Bfx} = \left(\frac{2 m_A}{m_A + m_B}\right) v_{Aix} + \left(\frac{m_B - m_A}{m_A + m_B}\right) v_{Bix}$$

For the case of equal masses (that is, $m_1 = m_2$), we find that $v_{Afx} = v_{Bix}$ and $v_{Bfx} = v_{Aix}$.
   Get Help: Picture It – Momentum Conservation

7.25 A (Block A rises to a height greater than H_A and block B rises to a height less than H_B).

Conservation of momentum:
$$\vec{P}_f = \vec{P}_i$$
$$m_A \vec{v}_{Af} + m_B \vec{v}_{Bf} = m_A \vec{v}_{Ai} + m_B \vec{v}_{Bi}$$
$$m_A v_{Afx} + m_B v_{Bfx} = m_A \sqrt{2gH_A} - m_B \sqrt{2gH_B}$$

Conservation of energy:
$$K_f = K_i$$
$$\frac{1}{2} m_A v_{Afx}^2 + \frac{1}{2} m_B v_{Bfx}^2 = \frac{1}{2} m_A v_{Aiy}^2 + \frac{1}{2} m_B v_{Biy}^2$$
$$m_A v_{Afx}^2 + m_B v_{Bfx}^2 = m_A(2gH_A) + m_B(2gH_B)$$

Solving this system of equations for the final speeds (that is, the speeds just after the collision) and then applying conservation of energy to calculate the final heights, we'll find that the less massive block (block A) will rise to a greater height than the more massive block (block B).

We can also look at the extremes of the system. What if block B is way more massive than block A? If they are released from about the same heights, then certainly block A rises higher after the collision. If they are released so that block B is nearly at the bottom initially, then it's going slowly when they collide, perhaps at nearly zero speed, but that, too, makes block A rise higher. What if block B is just slightly more massive than block A? If they are released from about the same height, then they have about the same speed when they collide, but that would result in block A having just a slightly higher speed afterward. So again, block A rises higher after the collision. Also, if block B is just slightly more massive than block A and block B is released near the bottom, block B is at close to zero velocity when the two blocks collide. Since block A is less massive than block B, it rebounds at a slightly higher speed than its initial speed. This means block A rises higher.

Get Help: Picture It – Momentum Conservation

Estimation/Numerical Analysis

7.29 Roughly speaking, it is around your navel. The ratio of the center of mass to height in humans is about 0.55.

Get Help: Picture It – Center of Mass

7.33 A tennis ball has a mass of 0.057 kg and a coefficient of restitution of 0.7. If an incoming volley is 40 m/s, then we get an impulse of around 4 kg · m/s.

$$\text{coefficient of restitution} = e = -\frac{v_{fx}}{v_{ix}}$$

$$v_{fx} = -ev_{ix} = -0.7v_{ix}$$

$$\vec{F}_{collision}\Delta t = \vec{p}_f - \vec{p}_i = m\vec{v}_f - m\vec{v}_i$$

$$F_{collision,x}\Delta t = m(v_{fx} - v_{ix}) = -m(0.7v_{ix} + v_{ix})$$

$$= -(0.057 \text{ kg})(1.7)(40 \text{ m/s}) \approx 4\frac{\text{kg}\cdot\text{m}}{\text{s}}$$

Get Help: P'Cast 7.9 – Follow-through

Problems

7.37

SET UP

The magnitude of the instantaneous momentum of a 0.057-kg tennis ball is 2.6 kg · m/s. We can use the definition of momentum to calculate the instantaneous speed of the ball.

SOLVE

$$p_x = mv_x$$

$$v_x = \frac{p_x}{m} = \frac{\left(2.6\frac{\text{kg}\cdot\text{m}}{\text{s}}\right)}{0.057 \text{ kg}} = \boxed{46\frac{\text{m}}{\text{s}}}$$

REFLECT

We would expect a large speed since the mass of the ball is small. This is a reasonable speed for a tennis ball.

Get Help: P'Cast 7.1 – Momentum and Kinetic Energy

7.41

SET UP

A 55.0-kg girl is riding a skateboard at a speed of 6.00 m/s. We can calculate the magnitude of her momentum directly from the definition of momentum. We are given the magnitude of the skateboard's momentum and we know that it must be traveling at the same speed as the girl. Therefore, we can find the mass of the skateboard through division.

SOLVE

Part a)

$$p_x = mv_x = (55.0 \text{ kg})\left(6.00\frac{\text{m}}{\text{s}}\right) = \boxed{3.30 \times 10^2 \frac{\text{kg}\cdot\text{m}}{\text{s}}}$$

Part b)

$$m = \frac{p_x}{v_x} = \frac{\left(30.0\frac{\text{kg}\cdot\text{m}}{\text{s}}\right)}{\left(6.00\frac{\text{m}}{\text{s}}\right)} = \boxed{5.00 \text{ kg}}$$

REFLECT

The momentum of the skateboard is one-eleventh of the momentum of the girl, which means the mass of the skateboard will be one-eleventh of the mass of the girl.

Get Help: P'Cast 7.1 – Momentum and Kinetic Energy

7.45

SET UP

An object of mass $3M$ has an initial velocity of v_0 in the x direction. It breaks into two unequal pieces. Piece 1 has a mass of M and travels off at an angle of $45°$ below the x-axis with a speed of v_1. Piece 2 has a mass of $2M$ and travels off at an angle of $30°$ above the x-axis with a speed of v_2. We can use conservation of momentum in order to calculate the final velocities of the two pieces. Since this is a two-dimensional problem, we will need to split the momenta into components and solve the x and y component equations.

Figure 7-1 Problem 45

SOLVE

y component:

$$P_{iy} = P_{fy}$$

$$P_{iy} = p_{2fy} + p_{1fy}$$

$$0 = (2M)v_2\sin(30°) - (M)v_1\sin(45°)$$

$$v_2 = \frac{v_1\sin(45°)}{2\sin(30°)} = \frac{v_1}{\sqrt{2}}$$

x component:

$$P_{ix} = P_{fx}$$

$$P_{ix} = p_{2fx} + p_{1fx}$$

$$(3M)v_0 = (2M)v_2\cos(30°) + (M)v_1\cos(45°)$$

$$3v_0 = 2v_2\cos(30°) + v_1\cos(45°) = 2\left(\frac{v_1}{\sqrt{2}}\right)\left(\frac{\sqrt{3}}{2}\right) + \frac{v_1}{\sqrt{2}}$$

$$v_1 = \frac{3\sqrt{2}}{\sqrt{3}+1} v_0 = 1.55 v_0$$

$$v_2 = \frac{v_1}{\sqrt{2}} = \frac{3}{\sqrt{3}+1} v_0 = 1.10 v_0$$

Therefore,

$$v_{1x} = 1.55 v_0 \cos 45° = 1.10 v_0$$

$$v_{1y} = -1.55 v_0 \sin 45° = -1.10 v_0$$

$$v_{2x} = 1.10 v_0 \cos 30° = 0.953 v_0$$

$$v_{2y} = 1.10 v_0 \sin 30° = 0.550 v_0$$

REFLECT

Initially there is no momentum in the y direction, which means the final y momentum must also be zero. If we calculate it explicitly, we see that this is true: $(M)(-1.10 v_0) + (2M)(0.550 v_0) = 0$.

Get Help: P'Cast 7.2 – Conservation of Momentum: A Collision on the Ice

7.47

SET UP

Two bighorn sheep are in the midst of a head-butting contest. Sheep A(m_A = 95.0 kg) moves at a speed of 10.0 m/s directly toward sheep B (m_B = 80.0 kg) running at 12.0 m/s. To determine which sheep wins the contest, we need to compare the momenta of the two sheep; whichever sheep has the larger momentum will end up knocking the other sheep backward.

SOLVE

Sheep A:

$$p_{Ax} = m_A v_{Ax} = (95.0 \text{ kg})\left(10.0 \frac{\text{m}}{\text{s}}\right) = 950 \frac{\text{kg} \cdot \text{m}}{\text{s}} = 9.5 \times 10^2 \frac{\text{kg} \cdot \text{m}}{\text{s}}$$

Sheep B:

$$p_{Bx} = m_B v_{Bx} = (80.0 \text{ kg})\left(-12.0 \frac{\text{m}}{\text{s}}\right) = -960 \frac{\text{kg} \cdot \text{m}}{\text{s}} = -9.6 \times 10^2 \frac{\text{kg} \cdot \text{m}}{\text{s}}$$

The lighter sheep has a larger momentum, so $\boxed{\text{sheep B}}$ will win the head-butting contest.

REFLECT

If we knew how long the sheep were in contact, we could calculate the force of one sheep on the other. We could then relate this to the acceleration of the sheep via Newton's second law. If the net force on the sheep were constant, we could use the constant acceleration equations to calculate how far sheep B pushes sheep A.

Get Help: P'Cast 7.2 – Conservation of Momentum: A Collision on the Ice

7.51

SET UP

A howler monkey (m_A = 5.00 kg) is swinging from a vine at a speed of v_{Aix} = 12.0 m/s due east toward a second monkey (m_B = 6.00 kg) that is also moving east at a speed of v_{Bix} = 8.00 m/s. The monkeys grab onto one another and travel together on the same vine. We can calculate the final speed v_f of the two monkeys traveling together by applying conservation of momentum. In our coordinate system, east will point in the positive direction.

SOLVE

$$P_{ix} = P_{fx}$$

$$m_A v_{Aix} + m_B v_{Bix} = (m_A + m_B) v_{fx}$$

$$v_{fx} = \frac{m_A v_{Aix} + m_B v_{Bix}}{m_A + m_B} = \frac{(5.00 \text{ kg})\left(12.0 \frac{\text{m}}{\text{s}}\right) + (6.00 \text{ kg})\left(8.00 \frac{\text{m}}{\text{s}}\right)}{(5.00 \text{ kg}) + (6.00 \text{ kg})} = \boxed{9.82 \frac{\text{m}}{\text{s}}}$$

REFLECT

It makes sense that the monkeys will continue to travel east at a speed in between their initial speeds.

Get Help: Interactive Exercise – Bullet Penetrates Can
Interactive Exercise – Ballistic Pendulum

7.55

SET UP

A block of ice (m_1 = 10.0 kg) is moving at 8.00 m/s toward the east when it collides elastically with a second block of ice (m_2 = 6.00 kg) moving toward the east at 4.00 m/s. We can use conservation of momentum and conservation of mechanical energy to find the final velocities of the two blocks. We can derive a useful relationship between the final velocities and the initial velocities of a one-dimensional elastic collision: $v_{1fx} = \left(\frac{m_1 - m_2}{m_1 + m_2}\right) v_{1ix} + \left(\frac{2m_2}{m_1 + m_2}\right) v_{2ix}$ and $v_{2fx} = \left(\frac{2m_1}{m_1 + m_2}\right) v_{1ix} + \left(\frac{m_2 - m_1}{m_1 + m_2}\right) v_{2ix}$. In our coordinate system, we'll consider +x to point toward the east.

SOLVE

$$v_{1fx} = \left(\frac{m_1 - m_2}{m_1 + m_2}\right) v_{1ix} + \left(\frac{2m_2}{m_1 + m_2}\right) v_{2ix}$$

$$= \left(\frac{(10.0 \text{ kg}) - (6.00 \text{ kg})}{(10.0 \text{ kg}) + (6.00 \text{ kg})}\right)\left(8.00 \frac{\text{m}}{\text{s}}\right) + \left(\frac{2(6.00 \text{ kg})}{(10.0 \text{ kg}) + (6.00 \text{ kg})}\right)\left(4.00 \frac{\text{m}}{\text{s}}\right) = 5.00 \frac{\text{m}}{\text{s}}$$

Block 1's final velocity is $\boxed{5.00 \text{ m/s to the right}}$.

$$v_{2fx} = \left(\frac{2m_1}{m_1 + m_2}\right) v_{1ix} + \left(\frac{m_2 - m_1}{m_1 + m_2}\right) v_{2ix}$$

$$= \left(\frac{2(10.0 \text{ kg})}{(10.0 \text{ kg}) + (6.00 \text{ kg})}\right)\left(8.00\frac{\text{m}}{\text{s}}\right) + \left(\frac{(6.00 \text{ kg}) - (10.0 \text{ kg})}{(10.0 \text{ kg}) + (6.00 \text{ kg})}\right)\left(4.00\frac{\text{m}}{\text{s}}\right) = 9.00\frac{\text{m}}{\text{s}}$$

Block 2's final velocity is $\boxed{9.00 \text{ m/s to the right}}$.

REFLECT

It makes sense that block 1 should slow down and block 2 should move faster after the collision.

Get Help: Interactive Exercise – Bullet Penetrates Can
Interactive Exercise – Ballistic Pendulum

7.59

SET UP

A 0.200-kg ball is initially traveling at 20.0 m/s toward your hand. You catch the ball and it comes to rest within 0.0250 s. We can find the magnitude of the force your hand exerts on the ball from the change in the ball's momentum and the contact time.

SOLVE

$$F_{\text{collision},x} = \left|\frac{\Delta p_x}{\Delta t}\right| = \left|\frac{m(v_{fx} - v_{ix})}{\Delta t}\right| = \left|\frac{(0.200 \text{ kg})\left(0 - \left(20.0\frac{\text{m}}{\text{s}}\right)\right)}{0.0250 \text{ s}}\right| = \boxed{1.60 \times 10^2 \text{ N}}$$

REFLECT

The force of your hand on the ball will act opposite to the ball's initial velocity.

Get Help: Picture It – Momentum Conservation

7.65

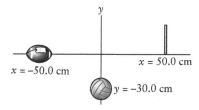

Figure 7-2 Problem 65

SET UP

The locations of three uniform objects (a rod, a football, and a volleyball) are given. Since the objects are uniform, the center of mass of each object is located at its geometrical center. The center of an ellipse is halfway along each axis. (Remember that the semimajor axis is the longer axis.)

SOLVE
Rod:

$$\boxed{\begin{array}{l} x_{\text{CM}} = 50.0 \text{ cm} \\ y_{\text{CM}} = 15.0 \text{ cm} \end{array}}$$

Football:

$$\boxed{\begin{aligned} x_{CM} &= -50.0 \text{ cm} \\ y_{CM} &= 0 \end{aligned}}$$

Volleyball:

$$\boxed{\begin{aligned} x_{CM} &= 0 \\ y_{CM} &= -30.0 \text{ cm} \end{aligned}}$$

REFLECT
Because we knew the dimensions of each shape and that they were uniform, we did not need to know the mass of the objects. We would get the same answers if we explicitly calculated the center of mass in each case.

Get Help: P'Cast 7.9 – Follow-through

General Problems

7.69

SET UP
Three hundred million people fall from a height of 1.00 m onto the ground. We can use conservation of mechanical energy to calculate the speed with which the people land on Earth. From this speed, we can calculate the magnitude of the momentum of all 300 million people; this is the momentum that will be imparted to Earth. Dividing this quantity by the mass of the Earth will give us the change in speed of the Earth.

SOLVE

Part a)
Speed of a single person:

$$mgh = \frac{1}{2}mv_{people,y}^2$$

$$v_{people,y} = \sqrt{2gh} = \sqrt{2\left(9.80\frac{m}{s^2}\right)(1.00 \text{ m})} = 4.43\frac{m}{s}$$

Momentum of 300 million people:

$$p_{people,y} = m_{people}v_{people,y} = (3.00 \times 10^8)(65.0 \text{ kg})\left(4.43\frac{m}{s}\right) = \boxed{8.64 \times 10^{10}\frac{kg \cdot m}{s}}$$

Part b)

$$\Delta p_{Earth,y} = m_{Earth}\Delta v_{Earth,y}$$

$$\Delta v_{Earth,y} = \frac{\Delta p_{Earth,y}}{m_{Earth}} = \frac{\left(8.64 \times 10^{10}\frac{kg \cdot m}{s}\right)}{5.98 \times 10^{24} \text{ kg}} = \boxed{1.44 \times 10^{-14}\frac{m}{s}}$$

REFLECT

We could have also used kinematics to calculate the landing speed of the people. Technically Earth will recoil at a nonzero speed, but a speed on the order of 10^{-14} m/s is imperceptible, so the Earth effectively stays still.

7.73

SET UP

A train car (m_{train} = 8000 kg) is rolling at an initial speed of v_{ix} = 20.0 m/s when, all of a sudden, rainwater collects inside of it. The final speed of the filled train car is v_{fx} = 19.0 m/s. We can calculate the mass of the rainwater from conservation of momentum.

SOLVE

$$m_{train}v_{ix} = (m_{train} + m_{water})v_{fx}$$

$$m_{water} = \frac{m_{train}v_{ix}}{v_{fx}} - m_{train} = \frac{(8000 \text{ kg})\left(20.0\frac{\text{m}}{\text{s}}\right)}{\left(19.0\frac{\text{m}}{\text{s}}\right)} - (8000 \text{ kg}) = \boxed{421 \text{ kg} = 400 \text{ kg}}$$

REFLECT

The speed of the train car decreased by about 5%, which means the mass must have increased by about 5%.

Get Help: Interactive Exercise – Bullet Penetrates Can
Interactive Exercise – Ballistic Pendulum

7.77

SET UP

A lion (m_L = 135 kg) is running northward at 80.0 km/h when it collides with and latches onto a gazelle (m_G = 29.0 kg) that is running eastward at 60.0 km/h. In our coordinate system east will point toward $+x$ and north will point toward $+y$. We can use conservation of momentum to find the final speed and direction of the lion–gazelle system after the lion attacks. Since this is a two-dimensional problem, we will need to split the momenta into components and solve the x and y component equations.

SOLVE

x component:

$$m_L v_{Lix} + m_G v_{Gix} = (m_L + m_G)v_{fx}$$

$$v_{fx} = \frac{m_L v_{Lix} + m_G v_{Gix}}{(m_L + m_G)} = \frac{0 + (29.0 \text{ kg})\left(60.0\frac{\text{km}}{\text{h}}\right)}{(135 \text{ kg}) + (29.0 \text{ kg})} = 10.6\frac{\text{km}}{\text{h}}$$

y component:

$$m_L v_{Liy} + m_G v_{Giy} = (m_L + m_G)v_{fy}$$

78 Chapter 7 Momentum, Collisions, and the Center of Mass

$$v_{fy} = \frac{m_L v_{Liy} + m_G v_{Giy}}{(m_L + m_G)} = \frac{(135 \text{ kg})\left(80.0 \frac{\text{km}}{\text{h}}\right) + 0}{(135 \text{ kg}) + (29.0 \text{ kg})} = 65.8 \frac{\text{km}}{\text{h}}$$

Final speed:

$$v = \sqrt{v_{fx}^2 + v_{fy}^2} = \sqrt{\left(10.6 \frac{\text{km}}{\text{h}}\right)^2 + \left(65.8 \frac{\text{km}}{\text{h}}\right)^2} = \boxed{66.6 \frac{\text{km}}{\text{h}}}$$

Final direction:

$$\theta = \tan^{-1}\left(\frac{65.8}{10.6}\right) = \boxed{80.8° \text{ north of east}}$$

REFLECT

The momentum of the lion is considerably larger than the momentum of the gazelle, so the final momentum of the system should point more north than east.

Get Help: Picture It – Momentum Conservation

7.79

SET UP

A bullet ($m_B = 0.0120$ kg) is shot into a block of wood of mass m_W at a speed of $v_{Bix} = 250$ m/s. The wood is attached to a spring with a spring constant $k = 200$ N/m. Once the bullet embeds itself into the wood, the spring compresses a distance of $x = 30.0$ cm before coming to a stop. We first need to use conservation of momentum in order to calculate the speed of the (bullet + wood) after the collision in terms of m_W. Then we can use conservation of mechanical energy to solve for the numerical value of m_W.

Figure 7-3 Problem 79

SOLVE
Conservation of momentum:

$$m_B v_{Bix} + m_W v_{Wix} = (m_B + m_W) v_{fx}$$

$$v_{fx} = \frac{m_B v_{Bix} + m_W v_{Wix}}{m_B + m_W} = \frac{m_B v_{Bix} + 0}{m_B + m_W} = \frac{m_B v_{Bix}}{m_B + m_W}$$

Conservation of mechanical energy:

$$U_i + K_i = U_f + K_f$$

$$0 + \frac{1}{2}(m_B + m_W) v_{fx}^2 = \frac{1}{2} k x^2 + 0$$

$$(m_B + m_W) v_{fx}^2 = k x^2$$

$$(m_B + m_W)\left(\frac{m_B v_{Bix}}{m_B + m_W}\right)^2 = kx^2$$

$$\frac{1}{(m_B + m_W)}(m_B v_{Bix})^2 = kx^2$$

$$m_W = \frac{(m_B v_{Bix})^2}{kx^2} - m_B = \frac{(0.0120 \text{ kg})^2 \left(250\frac{\text{m}}{\text{s}}\right)^2}{\left(200\frac{\text{N}}{\text{m}}\right)(0.300 \text{ m})^2} - (0.0120 \text{ kg}) = \boxed{0.488 \text{ kg}}$$

REFLECT

The block of wood with the embedded bullet moves at a speed of 6.00 m/s.

Get Help: Interactive Exercise – Bullet Penetrates Can
Interactive Exercise – Ballistic Pendulum

Chapter 8
Rotational Motion

Conceptual Questions

8.3 The moment of inertia of a rotating object is calculated relative to the axis of rotation. If we don't explicitly define this axis at the outset of the problem, the calculated moment of inertia will be ambiguous at best and meaningless at worst. Even for axes crossing the center of mass, different axes will have different moments.

   Get Help: Picture It – Moment of Inertia

8.7 Part a) When riding a seesaw, the torque produced by each rider about the pivot should be approximately equal in magnitude. The riders are trying to move their centers of mass relative to the pivot of the seesaw by leaning forward or backward. This, in turn, will affect the magnitude of the torque about the pivot produced by each rider by changing the moment arms.

Part b) We are told that the riders are sitting equidistant from the pivot point and the person at the top always leans backward, which means the person at the top has a smaller mass than the person at the bottom (who always leans forward).

8.13 If we consider the system to be the cookie dough + turntable, there is no net torque acting on the system, and the rotational kinetic energy remains constant since there is no net work done on the system. This means the angular momentum of the system is constant. Dropping the cookie dough onto the edge of the turntable changes the way the mass in the system is distributed about the axis of rotation, thus changing the moment of inertia. Since the angular momentum is constant and the moment of inertia of the system changes, the angular velocity of the system must change after the cookie dough lands. (The moment of inertia increases. Therefore, the magnitude of the angular velocity decreases; that is, the turntable slows down.) A changing angular velocity means there is a nonzero angular acceleration.

8.17 We can use conservation of energy to answer this question. All three objects—a sphere, a cylinder, and a ring—start at rest from the same height and, thus, have the same initial potential energy. As the objects roll down the inclined plane, potential energy is converted into rotational *and* translational kinetic energy. The smaller the moment of inertia of an object, the more energy that will be converted into translational kinetic energy rather than rotational kinetic energy. The object that reaches the bottom of the ramp has the largest translational kinetic energy, so the sphere $\left(I_{sphere} = \frac{2}{5}MR^2\right)$ "wins," followed by the cylinder $\left(I_{cylinder} = \frac{1}{2}MR^2\right)$, and then the hoop $(I_{hoop} = MR^2)$.

8.21 Counterclockwise corresponds to the positive sense of rotation. We're told the merry-go-round undergoes constant acceleration.

a) Graph (c). The angular speed increases linearly as a function of time.

b) Graph (i). The angular displacement increases parabolically under constant angular acceleration.

c) Graph (a). The merry-go-round maintains a constant maximum angular speed.

d) Graph (ii). The angular displacement increases linearly for a constant angular speed.

e) Graph (d). The angular speed should decrease linearly from a maximum to zero.

f) Graph (iv). The angular position of the merry-go-round will eventually level out once it comes to a stop.

g) When the merry-go-round is speeding up in the counterclockwise direction, there is a positive angular acceleration caused by a positive net torque on the merry-go-round. Since we're told that the angular acceleration is constant, the net torque must also be constant. The net torque is equal to zero when the merry-go-round is spinning at a constant angular velocity. The merry-go-round slows down due to a constant, negative net torque being applied.

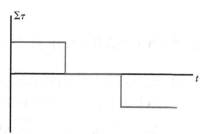

Figure 8-1 Problem 21

Get Help: Picture It – Angular Momentum

Multiple-Choice Questions

8.27 C (the hoop). Three objects—a solid ball, a solid disk, and a hoop—that have the same mass and radius start rolling *and* translating up an incline from the same height, which we'll call $y = 0$. The objects all begin moving at the same linear speed. The moment of inertia for each of the objects is different, though:

$I_{\text{solid sphere}} = \frac{2}{5}MR^2$, $I_{\text{solid disk}} = \frac{1}{2}MR^2$, $I_{\text{hoop}} = MR^2$.

We can use conservation of mechanical energy to solve this problem:

$$K_{\text{rotational,i}} + K_{\text{translational,i}} + U_i = K_{\text{rotational,f}} + K_{\text{translational,f}} + U_f$$

$$\frac{1}{2}I_{\text{CM}}\omega^2 + \frac{1}{2}Mv_{\text{CM}}^2 + 0 = 0 + 0 + Mgh$$

$$\frac{1}{2}I_{\text{CM}}\left(\frac{v_{\text{CM}}}{R}\right)^2 + \frac{1}{2}Mv_{\text{CM}}^2 = Mgh$$

Solving for the height h that each object makes it up the incline, $h = \frac{v_{\text{CM}}^2}{2Mg}\left(\frac{I}{R^2} + M\right)$.

Since the moment of inertia for the hoop has the largest coefficient (1 versus $\frac{2}{5}$ versus $\frac{1}{2}$), it will go up the incline the farthest.

8.31 B ($2MR^2$). We need to find the moment of inertia of a thin ring of radius R rotating about a point on its rim. The parallel-axis theorem will give us the moment of inertia of an object rotating about an axis *not* passing through its center of mass: $I = I_{CM} + Mh^2$, where I_{CM} is the moment of inertia of the object rotating about an axis through its center of mass and h is the distance between that axis and the actual rotation axis. For a thin ring, $I_{CM} = MR^2$. The distance from the center of mass to a point on the edge of the ring is R. Therefore, $I = I_{CM} + Mh^2 = MR^2 + MR^2 = 2MR^2$.

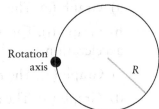

Figure 8-2 Problem 31

Get Help: Picture It – Moment of Inertia
P'Cast 8.4 – Moment of Inertia for Dumbbells I

Estimation/Numerical Analysis

8.33 We're told that the cloverleaf is approximately three-quarters of a circle, which corresponds to a total angular displacement of $\Delta\theta = \frac{3}{4}(2\pi \text{ rad}) = \frac{3\pi}{2}$rad. We need to estimate the amount of time the car is driving on the cloverleaf. Let's say that the cloverleaf is three-quarters of a circle of 25 m in radius, and the car is driving a little over 30 mph (15 m/s). The total distance the car travels is the arc length, or $\left(\frac{3\pi}{2}\text{rad}\right)(25 \text{ m}) = \frac{75\pi}{2}$m. It takes the car $\frac{\left(\frac{75\pi}{2}\text{m}\right)}{\left(15\frac{\text{m}}{\text{s}}\right)} = \frac{5\pi}{2}$s to drive around the cloverleaf. The angular speed is the angular displacement over the time it took, so $\omega_z = \frac{\Delta\theta}{\Delta t} = \frac{\left(\frac{3\pi}{2}\text{rad}\right)}{\left(\frac{5\pi}{2}\text{s}\right)} = \boxed{\frac{3}{5}\frac{\text{rad}}{\text{s}} = 0.6\frac{\text{rad}}{\text{s}}}$.

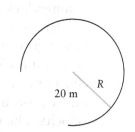

Figure 8-3 Problem 33

Get Help: P'Cast 8.1 – Speed vs. Angular Speed

8.41 The magnitude of the "skip-it" ball's angular momentum is equal to $L_z = I\omega_z = MR^2\omega_z$, where we plugged in the moment of inertia for a point mass. Since the toy is made of plastic and is attached to a child's leg, the mass of the ball should be around 0.25 kg. The toy is approximately a meter long. The ball makes about a revolution per second when the child is swinging it around her leg; this corresponds to an angular speed of $\omega_z = 1\frac{\text{rev}}{\text{s}} \times \frac{2\pi \text{ rad}}{1 \text{ rev}} = 2\pi\frac{\text{rad}}{\text{s}}$. Plugging all of this in yields $L_z = (0.25 \text{ kg})(1 \text{ m})^2 \times \left(2\pi\frac{\text{rad}}{\text{s}}\right) = \frac{\pi}{2}\frac{\text{kg} \cdot \text{m}^2}{\text{s}} \approx \boxed{2\frac{\text{kg} \cdot \text{m}^2}{\text{s}}}$.

Get Help: Picture It – Angular Momentum

Problems

8.47

SET UP

It is reasonable to assume that the angular speed of the Moon is roughly constant. In this case, the angular speed is given by the angular displacement divided by the time it took. We're told that the Moon completes one orbit (assuming a circular orbit) in 27.4 days. One circular orbit is an angular displacement of 2π radians.

SOLVE

$$1 \text{ orbit} = 27.4 \text{ d} \times \frac{24 \text{ h}}{1 \text{ d}} \times \frac{3600 \text{ s}}{1 \text{ h}} = 2.367 \times 10^6 \text{ s}$$

$$\omega_{\text{average},z} = \frac{\Delta\theta}{\Delta t} = \frac{2\pi \text{ rad}}{2.367 \times 10^6 \text{ s}} = \boxed{2.65 \times 10^{-6} \frac{\text{rad}}{\text{s}}}$$

REFLECT

Since it takes such a long time for the Moon to complete one orbit, we expect its angular speed to be a small number. The Earth–Moon distance is not necessary for finding the angular speed.

Get Help: P'Cast 8.1 – Speed vs. Angular Speed

8.51

SET UP

We are given the moment of inertia I and rotational kinetic energy $K_{\text{rotational}}$ of an object and asked to find its angular speed. We can use the definition of rotational kinetic energy, $K_{\text{rotational}} = \frac{1}{2}I\omega^2$, and solve for ω. To convert from radians per second, we need to realize that one revolution corresponds to an angular displacement of 2π radians.

SOLVE

Solving for ω and plugging in the given values, we find that

$$K_{\text{rotational}} = \frac{1}{2}I\omega^2$$

$$\omega = \sqrt{\frac{2K_{\text{rotational}}}{I}} = \sqrt{\frac{2(2.75 \text{ J})}{0.330 \text{ kg} \cdot \text{m}^2}} = \boxed{4.08 \frac{\text{rad}}{\text{s}}}$$

$$= \left(4.08 \frac{\text{rad}}{\text{s}}\right) \times \frac{60 \text{ s}}{1 \text{ min}} \times \frac{1 \text{ rev}}{2\pi \text{ rad}} = \boxed{39.0 \frac{\text{rev}}{\text{min}}}$$

REFLECT

Since we are given the moment of inertia explicitly, we do not need to use the fact that it is a rotating wheel.

Get Help: P'Cast 8.2 – Turbine Kinetic Energy

8.55

SET UP

The steering wheel is rotating about an axis that passes through its center and is made up of a rim of radius R and mass M and five thin rods of length R and mass $M/2$. The overall moment of inertia of the steering wheel is equal to the moment of inertia of the rim plus the moments of inertia for each of the thin rods. We can model the rim as a uniform ring of radius R rotating about an axis through its center, and each of the thin rods is rotating about a perpendicular line through one end.

Figure 8-4 Problem 55

SOLVE

$$I_{\text{rim}} = I_{\text{ring}} = MR^2$$

$$I_{\text{rod}} = \frac{1}{3}\left(\frac{M}{2}\right)R^2 = \frac{1}{6}MR^2$$

$$I_{\text{total}} = I_{\text{rim}} + 5I_{\text{rod}} = MR^2 + 5\left(\frac{1}{6}MR^2\right) = MR^2 + \frac{5}{6}MR^2 = \boxed{\frac{11}{6}MR^2}$$

REFLECT

Since the steering wheel is made up of simple geometric objects, we can use the results in Table 8-1 of the textbook. We then determine the axis of rotation for the overall object. This allows us to determine which moments of inertia to use (for example, a thin rod rotating about an axis through its center versus a thin rod rotating about an axis through one end).

Get Help: Picture It – Moment of Inertia
P'Cast 8.4 – Moment of Inertia for Dumbbells I

8.61

SET UP

The system is made up of three separate objects: a solid sphere of radius R and mass M, another solid sphere of radius $2R$ and mass M, and a thin, uniform rod of length $3R$ and mass M. We need to find the moment of inertia about the axis through the center of the rod by considering the moments of inertia for each of the component objects in order to find the overall moment of inertia of the system. Since the two spheres are not located at the axis of rotation, we need to determine the distance their centers of mass are from the axis of rotation and invoke the parallel-axis theorem. The center of mass for the large sphere is $3.5R$ from the axis of rotation, and the small sphere is $2.5R$ from the axis.

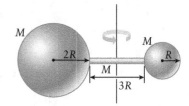

Figure 8-5 Problem 61

SOLVE

$$I_{total} = I_{rod} + I_{left\ sphere} + I_{right\ sphere}$$

$$I_{rod} = \frac{1}{12}M(3R)^2 = \frac{9}{12}MR^2$$

$$I_{left\ sphere} = I_{sphere} + I_{from\ axis} = \frac{2}{5}M(2R)^2 + M\left(2R + \frac{3}{2}R\right)^2 = \frac{8}{5}MR^2 + \frac{49}{4}MR^2 = \frac{277}{20}MR^2$$

$$I_{right\ sphere} = I_{sphere} + I_{from\ axis} = \frac{2}{5}MR^2 + M\left(R + \frac{3}{2}R\right)^2 = \frac{2}{5}MR^2 + \frac{25}{4}MR^2 = \frac{133}{20}MR^2$$

$$I_{total} = I_{rod} + I_{left\ sphere} + I_{right\ sphere} = \frac{9}{12}MR^2 + \frac{277}{20}MR^2 + \frac{133}{20}MR^2 = \boxed{\frac{85}{4}MR^2}$$

REFLECT

The system is made up of simple geometric objects, so we can use the results in Table 8-1 of the textbook. We need to be careful to add the distance the center of mass of each sphere is from the rotation axis in determining the overall moment of inertia. Since the solid spheres are not located along the axis of rotation, it makes sense that the overall moment of inertia of the system is larger than just the sum of

$$I_{rod} + I_{sphere\ of\ 2R} + I_{sphere\ of\ R} = \frac{9}{12}MR^2 + \frac{8}{5}MR^2 + \frac{2}{5}MR^2 = \frac{33}{12}MR^2$$

Get Help: Picture It – Moment of Inertia
P'Cast 8.7 – Using the Parallel-Axis Theorem II
P'Cast 8.8 – Earring Moment of Inertia

8.67

SET UP

A bowling ball of known mass $m = 5.00$ kg and radius $R = 11.0$ cm is rolling without slipping down a lane at an angular speed of $\omega = 2.00\frac{rad}{s}$. We need to determine the ratio of the ball's translational kinetic energy $\left(K_{translational} = \frac{1}{2}mv_{CM}^2\right)$ to its rotational kinetic energy $\left(K_{rotational} = \frac{1}{2}I_{CM}\omega^2\right)$. The translational speed of the ball is related to the angular speed by $v_{CM} = R\omega$. The ball is a sphere, so its moment of inertia is $I_{CM} = \frac{2}{5}mR^2$.

SOLVE

$$\frac{K_{translational}}{K_{rotational}} = \frac{\left(\frac{1}{2}mv_{CM}^2\right)}{\left(\frac{1}{2}I_{CM}\omega^2\right)} = \frac{m(R\omega)^2}{\left(\frac{2}{5}mR^2\right)\omega^2} = \boxed{\frac{5}{2}}$$

REFLECT

The solution is *independent* of the mass of the ball, the size of the ball, or the speed the ball is rolling!

8.71

SET UP

A Frisbee of mass $m = 0.160$ kg and diameter $d = 0.250$ m is spinning at an angular speed of 300 rpm about an axis through its center. We are asked to find the rotational kinetic energy of the Frisbee. We are given the angular speed of the Frisbee but not the moment of inertia. The mass is *not* uniformly distributed in this case—70% of the mass can be modeled as a thin ring of diameter d, while the remaining 30% of the mass can be modeled as a flat disk of diameter d. The moment of inertia of the Frisbee is the sum of the moments of inertia for the ring and the disk.

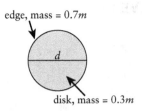

Figure 8-6 Problem 71

SOLVE

Kinetic energy of the Frisbee:

$$I_{\text{Frisbee}} = I_{\text{ring}} + I_{\text{disk}} = m_{\text{ring}}r_{\text{ring}}^2 + \frac{1}{2}m_{\text{disk}}r_{\text{disk}}^2 = (0.7m)\left(\frac{d}{2}\right)^2 + \frac{1}{2}(0.3m)\left(\frac{d}{2}\right)^2$$

$$= (0.7)(0.160 \text{ kg})\left(\frac{0.250 \text{ m}}{2}\right)^2 + \frac{1}{2}(0.3)(0.160 \text{ kg})\left(\frac{0.250 \text{ m}}{2}\right)^2 = 0.002125 \text{ kg} \cdot \text{m}^2$$

$$K_{\text{rotational}} = \frac{1}{2}I\omega^2 = \frac{1}{2}(0.002125 \text{ kg} \cdot \text{m}^2)\left(300\frac{\text{rev}}{\text{min}} \times \frac{1 \text{ min}}{60 \text{ s}} \times \frac{2\pi \text{ rad}}{1 \text{ rev}}\right)^2 = \boxed{1 \text{ J}}$$

REFLECT

If the mass of the Frisbee were distributed evenly, the rotational kinetic energy would be 0.62 J, which is approximately half of the calculated energy. This makes sense because the rotational inertia of a solid disk is one-half the rotational inertia of a ring.

8.77

SET UP

A CD player initially rotates at 300 rpm. It speeds up to 450 rpm within 0.75 s. We will assume that the CD player undergoes constant acceleration from 300 rpm to 450 rpm and we will use rotational kinematics. We will need to convert from rpm to rad/s.

SOLVE

$$450\frac{\text{rev}}{\text{min}} \times \frac{1 \text{ min}}{60 \text{ s}} \times \frac{2\pi \text{ rad}}{1 \text{ rev}} = 15\pi\frac{\text{rad}}{\text{s}}$$

$$300\frac{\text{rev}}{\text{min}} \times \frac{1 \text{ min}}{60 \text{ s}} \times \frac{2\pi \text{ rad}}{1 \text{ rev}} = 10\pi\frac{\text{rad}}{\text{s}}$$

$$\omega_z = \omega_{0z} + \alpha_z t$$

$$\alpha_z = \frac{\omega_z - \omega_{0z}}{t} = \frac{\left(15\pi\frac{\text{rad}}{\text{s}}\right) - \left(10\pi\frac{\text{rad}}{\text{s}}\right)}{0.75 \text{ s}} = \frac{20\pi}{3}\frac{\text{rad}}{\text{s}^2} = \boxed{21\frac{\text{rad}}{\text{s}^2}}$$

REFLECT

Although this seems large, it is reasonable given that the increase of 150 rpm happens in under a second. Remember to convert rpm into rad/s using 1 minute = 60 seconds and 1 revolution = 2π radians.

8.79

SET UP

You are holding a barbell of mass $m = 10$ kg in a completely outstretched hand of length $L = 0.75$ m. We're asked to calculate the magnitude of the torque due to the barbell about your shoulder, $\tau = rF\sin(\phi)$. The force of interest is the force of the barbell on your hand, which acts downward. We need to draw a free-body diagram of the barbell at rest and use Newton's second law in order to determine the magnitude of this force. The force of your arm on the barbell is related to the force of the barbell on your arm through Newton's third law. The distance from the rotation axis to the point where the force is applied is the length of the arm, L. Since the force of the barbell on the arm acts straight down and the arm is completely outstretched, the angle between the force vector and the $\vec{r}$ vector is 90°.

SOLVE

Free-body diagram of the barbell and Newton's second law:

$$\sum F_{ext,y} = F_{\text{arm on barbell}} - w = ma_y = 0$$

$$F_{\text{arm on barbell}} = w = mg$$

Newton's third law:

$$F_{\text{arm on barbell}} = F_{\text{barbell on arm}} = mg$$

Solving for the magnitude of the torque,

$$\tau = rF\sin(\phi) = Lmg\sin(90°) = (0.75 \text{ m})(10 \text{ kg})\left(9.80 \frac{\text{m}}{\text{s}^2}\right) = \boxed{74 \text{ N} \cdot \text{m}}$$

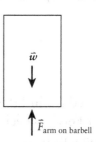

Figure 8-7 Problem 79

REFLECT

Since we're interested in the forces acting on the arm, we use the contact force of the barbell on the arm, not the *weight* of the barbell, in calculating the torque. The weight of the barbell is the force on the barbell due to the Earth, which would not appear on a free-body diagram of the arm and is *not* the third law partner of the contact force. The two forces are related, though; we just need to draw a free-body diagram for the barbell at rest and solve Newton's second law.

Get Help: Interactive Exercise – Sign

8.83

SET UP

A 75.0-N force acts on an 85.0-cm-wide door at various angles in parts (a)–(d). We are asked to calculate the torque that this force exerts about an axis through the hinges in each case. Because the force is exerted in the same spot on the door, the magnitude of $\vec{r}$ is the same in each case, namely, 85.0 cm. The angle ϕ between $\vec{r}$ and $\vec{F}$ changes, though.

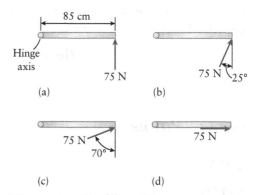

Figure 8-8 Problem 83

88 Chapter 8 Rotational Motion

SOLVE

Part a) $\phi = 90°$, so $\tau = rF\sin(\phi) = (0.850 \text{ m})(75.0 \text{ N})\sin(90°) = \boxed{63.8 \text{ N} \cdot \text{m}}$.

Part b) $\phi = 115°$, so $\tau = rF\sin(\phi) = (0.850 \text{ m})(75.0 \text{ N})\sin(115°) = \boxed{57.8 \text{ N} \cdot \text{m}}$.

Part c) $\phi = 160°$, so $\tau = rF\sin(\phi) = (0.850 \text{ m})(75.0 \text{ N})\sin(160°) = \boxed{21.8 \text{ N} \cdot \text{m}}$.

Part d) $\phi = 0°$, so $\tau = rF\sin(\phi) = (0.850 \text{ m})(75.0 \text{ N})\sin(0°) = \boxed{0}$.

REFLECT

The maximum torque should occur when the force is applied perpendicularly to the door (part a), and the minimum torque should occur when the force is applied parallel to the door (part d). For angles in between, we expect the torque to have an intermediate value.

Get Help: Interactive Exercise – Sign

8.97

SET UP

We are asked to find the magnitude of the angular momentum ($L = I\omega$) of a 70.0-kg person riding in a Ferris wheel of diameter $d = 35.0$ m that makes a complete revolution in 25.0 s. The person can be treated as a point mass rotating at a distance $d/2$ (that is, the radius) about the center of the Ferris wheel. We can calculate the angular speed of the person by knowing that 1 revolution is 2π radians and assuming the angular speed is constant.

SOLVE

$$I_{\text{point mass}} = mr^2 = m\left(\frac{d}{2}\right)^2 = (70.0 \text{ kg})\left(\frac{35.0 \text{ m}}{2}\right)^2 = 21{,}438 \text{ kg} \cdot \text{m}^2$$

$$\omega = \frac{\Delta\theta}{\Delta t} = \frac{2\pi \text{ rad}}{25.0 \text{ s}} = 0.251 \frac{\text{rad}}{\text{s}}$$

$$L = I\omega = (21{,}438 \text{ kg} \cdot \text{m}^2)\left(0.251 \frac{\text{rad}}{\text{s}}\right) = \boxed{5.39 \times 10^3 \frac{\text{kg} \cdot \text{m}^2}{\text{s}}}$$

REFLECT

Recall that r for the moment of inertia of a point mass is the distance the mass is located from the axis of rotation. In this case it is equal to the radius of the Ferris wheel since the passenger cars are located along the rim.

Get Help: Picture It – Angular Momentum
Interactive Exercise – Block

General Problems

8.105

SET UP

The bike rim is to be made up of a hoop of mass $m_{\text{hoop}} = 1.00$ kg and radius $R = 0.500$ m and an unknown number of spokes, each of mass $m_{\text{spoke}} = 0.0100$ kg. Since the spokes connect the center of a typical bicycle wheel to the hoop, the length of each one is equal to the radius of the hoop. The overall moment of inertia of the bike rim rotating about an axis through its center is equal to the moment of inertia of the hoop plus the moment of inertia for *each* spoke. We can model each spoke as a thin rod of length R rotating about an axis through one end. From this we can calculate the number of spokes necessary to give the stated overall moment of inertia. Once we know the number of spokes, we can determine the mass of all of the spokes and add that to the mass of the hoop to get the overall mass of the rim.

SOLVE

Part a)

$$I_{rim} = I_{hoop} + NI_{spoke} = m_{hoop}R^2 + N\left(\frac{1}{3}m_{spoke}L^2\right) = m_{hoop}R^2 + N\left(\frac{1}{3}m_{spoke}R^2\right)$$

$$N = (I_{rim} - m_{hoop}R^2)\left(\frac{3}{m_{spoke}R^2}\right)$$

$$= ((0.280 \text{ kg} \cdot \text{m}^2) - (1.00 \text{ kg})(0.500 \text{ m})^2)\left(\frac{3}{(0.0100 \text{ kg})(0.500 \text{ m})^2}\right) = \boxed{36.0}$$

Part b)

$$m_{rim} = m_{hoop} + Nm_{spoke} = (1.00 \text{ kg}) + 36.0(0.0100 \text{ kg}) = \boxed{1.36 \text{ kg}}$$

REFLECT

This is a reasonable mass for a bike rim. We aren't explicitly given the length of the spokes, but from our everyday knowledge about bicycle wheels we are able to get started on the problem. Picturing how a typical bike wheel is constructed and how it rotates helps determine which moment of inertia formulas are appropriate to use. Using everyday knowledge to make reasonable assumptions is an invaluable skill in solving physics problems.

Get Help: Picture It – Moment of Inertia

8.113

SET UP

When the space debris accumulates on the surface of the Earth, it changes both the mass and moment of inertia of the Earth, which is why we expect the Earth's angular velocity to decrease. We can use conservation of angular momentum if we treat the space debris and the Earth as an isolated system. Our initial state will be the Earth spinning at its "normal" rate (1 revolution in 24 h) as the space debris falls radially inward from space. We assume the debris falls radially inward until it hits the Earth's surface because this simplifies the calculation since the initial angular momentum of the space debris will be zero. The final state will be the Earth and space debris spinning together. We will assume the debris is uniformly distributed on the surface of the Earth in a thin layer. This allows us to model the final moment of inertia as a thin, spherical shell with the radius of the Earth plus the moment of inertia of the Earth itself, $I_f = I_{Earth} + I_{debris} = I_{solid\ sphere} + I_{spherical\ shell}$. Part (a) of the question asks for the *change* in the angular velocity per year, so we'll be solving for $\Delta\omega_z = \omega_{fz} - \omega_{iz}$. Once we know $\Delta\omega_z$ for one year we can use the definition of the period to determine the total change in angular speed if the Earth's rotation period changed by 1 s. Comparing these two values of $\Delta\omega_z$ will tell us how many years of accumulating debris it would take for that change to occur.

Chapter 8 Rotational Motion

SOLVE

Part a)

$$L_{iz} = L_{fz}$$

$$I_i \omega_{iz} = I_f \omega_{fz}$$

$$\omega_{fz} = \frac{I_i \omega_{iz}}{I_f}$$

$$\Delta \omega_z = \omega_{fz} - \omega_{iz} = \left(\frac{I_i \omega_{iz}}{I_f}\right) - \omega_{iz} = \omega_{iz}\left(\frac{I_i}{I_f} - 1\right) = \omega_{iz}\left(\frac{\left(\frac{2}{5}m_E R_E^2\right)}{\left(\frac{2}{5}m_E R_E^2\right) + \left(\frac{2}{3}m_{debris} R_E^2\right)} - 1\right)$$

$$= \omega_{iz}\left(\frac{m_E}{m_E + \left(\frac{5}{3}m_{debris}\right)} - 1\right) = \omega_{iz}\left(\frac{m_E}{m_E + \left(\frac{5}{3}m_{debris}\right)} - \frac{m_E + \left(\frac{5}{3}m_{debris}\right)}{m_E + \left(\frac{5}{3}m_{debris}\right)}\right)$$

$$= -\omega_{iz}\left(\frac{\left(\frac{5}{3}m_{debris}\right)}{m_E + \left(\frac{5}{3}m_{debris}\right)}\right)$$

Converting the angular velocity and mass of the debris:

$$\omega_{iz} = \frac{2\pi \text{ rad}}{24 \text{ h}} \times \frac{1 \text{ h}}{3600 \text{ s}} = 7.3 \times 10^{-5} \frac{\text{rad}}{\text{s}}$$

$$m_{debris} = 60{,}000 \text{ tons} \times \frac{2000 \text{ lb}}{1 \text{ ton}} \times \frac{1 \text{ kg}}{2.2 \text{ lb}} = 5.5 \times 10^7 \text{ kg}$$

Plugging in:

$$\Delta \omega_z = -\left(7.3 \times 10^{-5} \frac{\text{rad}}{\text{s}}\right)\left(\frac{\frac{5}{3}(5.5 \times 10^7 \text{ kg})}{(5.97 \times 10^{24} \text{ kg}) + \frac{5}{3}(5.5 \times 10^7 \text{ kg})}\right) = \boxed{-1 \times 10^{-21} \frac{\text{rad}}{\text{s}}}$$

This is the change in angular velocity in one year.

Part b)

$$T = \frac{2\pi}{\omega}$$

$$\Delta T = \frac{2\pi}{\omega_f} - \frac{2\pi}{\omega_i} = \frac{2\pi}{\omega_i + \Delta\omega} - \frac{2\pi}{\omega_i} = 2\pi\left(\frac{1}{\omega_i + \Delta\omega} - \frac{1}{\omega_i}\right)$$

$$\frac{\Delta T}{2\pi} + \frac{1}{\omega_i} = \frac{1}{\omega_i + \Delta\omega}$$

$$\omega_i + \Delta\omega = \frac{1}{\left(\frac{\Delta T}{2\pi} + \frac{1}{\omega_i}\right)} = \frac{1}{\left(\frac{\omega_i(\Delta T) + 2\pi}{2\pi\omega_i}\right)} = \frac{2\pi\omega_i}{\omega_i(\Delta T) + 2\pi}$$

$$\Delta\omega = \frac{2\pi\omega_i}{\omega_i(\Delta T) + 2\pi} - \omega_i = \omega_i\left(\frac{2\pi}{\omega_i(\Delta T) + 2\pi} - 1\right)$$

$$= \left(7.3 \times 10^{-5}\frac{\text{rad}}{\text{s}}\right)\left(\frac{2\pi}{\left(7.3 \times 10^{-5}\frac{\text{rad}}{\text{s}}\right)(1\text{ s}) + 2\pi} - 1\right)$$

$$= -8.5 \times 10^{-10}\frac{\text{rad}}{\text{s}}$$

This is the overall change in angular speed. To find how long it takes to reach this overall change in angular velocity, we need to divide it by the change in angular velocity *per year* (Note that the angular speed per year is the magnitude of the angular velocity calculated in Part a.):

$$\text{time} = \frac{\left(-8.5 \times 10^{-10}\frac{\text{rad}}{\text{s}}\right)}{\left(-1 \times 10^{-21}\frac{\text{rad}}{\text{s}\cdot\text{y}}\right)} = \boxed{8 \times 10^{11}\text{ y}}$$

REFLECT

Although 60,000 tons of debris sounds like a lot, it is negligible compared to the mass of the Earth. The amount of time we calculated for the Earth's rotation to change by a full second is longer than the age of the Earth (approximately 4×10^9 years), so we don't have to worry about space debris changing the Earth's rotation significantly...which is good.

Get Help: Picture It – Angular Momentum
Interactive Exercise – Block

8.119

SET UP

A marble of mass M, radius R, and moment of inertia $I = \frac{2}{5}MR^2$ is placed in front of a spring with spring constant k that has been compressed an amount x_c. The spring is released and, when the spring reaches its equilibrium length,

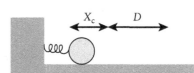

Figure 8-9 Problem 119

the marble comes off the spring and begins to roll without slipping. The statement that static friction does no work suggests that we use conservation of mechanical energy to determine the time t it takes for the marble to travel a distance D. The initial elastic potential energy stored in the spring is converted into both translational and rotational kinetic energy of the marble. By solving for the linear speed of the marble as it leaves the spring, we can then use the definition of speed to determine the time it takes to travel a distance D. In part (b) we can use dimensional analysis to ensure our algebraic answer from part (a) has the correct dimension of time.

SOLVE

Part a)
Conservation of mechanical energy:

$$U_{elastic} = K_{rotational} + K_{translational}$$

$$\frac{1}{2}kx_c^2 = \frac{1}{2}I\omega^2 + \frac{1}{2}Mv_{CM}^2$$

$$\frac{1}{2}kx_c^2 = \frac{1}{2}\left(\frac{2}{5}MR^2\right)\left(\frac{v_{CM}}{R}\right)^2 + \frac{1}{2}Mv_{CM}^2$$

$$\frac{1}{2}kx_c^2 = \frac{1}{5}Mv_{CM}^2 + \frac{1}{2}Mv_{CM}^2 = \frac{7}{10}Mv_{CM}^2$$

$$v_{CM} = \sqrt{\frac{5kx_c^2}{7M}} = \frac{D}{t}$$

$$t = \frac{D}{v_{CM}} = \frac{D}{\sqrt{\frac{5kx_c^2}{7M}}} = D\sqrt{\frac{7M}{5kx_c^2}} = \boxed{\frac{D}{x_c}\sqrt{\frac{7M}{5k}}}$$

Part b)

$$[t] = [T]$$

$$[M] = [M]$$

$$[D] = [L]$$

$$[x_c] = [L]$$

$$[k] = [M][T]^{-2}$$

$$[t] \stackrel{?}{=} \frac{[D]}{[x_c]}\sqrt{\frac{[M]}{[k]}}$$

$$[T] = \frac{[L]}{[L]}\sqrt{\frac{[M]}{[M][T]^{-2}}} = \sqrt{[T]^2} = [T]$$

REFLECT

Because the work done by all the nonconservative forces on the marble is equal to zero, we could invoke the conservation of mechanical energy because $W_{nonconservative} = \Delta E_{mechanical}$. We also explicitly use dimensional analysis in this problem. It's always a good idea to check the dimensions of your final answer whether or not it is required to do so.

Chapter 9
Elastic Properties of Matter: Stress and Strain

Conceptual Questions

9.3 Yes. Real cables have weight, and that weight can be sufficient to break the cable.

9.13 The shear modulus is a measure of how much an object will deform under a given stress. The larger the shear modulus, the less it will deform; it is more rigid. So, the term *rigidity* applies because a material will deform less if it has a larger shear modulus.

Multiple-Choice Questions

9.21 C ($\sqrt{2}$). The applied force, Young's modulus, and original length of the cable remain constant:

$$A_1 \Delta L_1 = A_2 \Delta L_2$$

$$\pi \left(\frac{D_1}{2}\right)^2 (\Delta L) = \pi \left(\frac{D_2}{2}\right)^2 \left(\frac{\Delta L}{2}\right)$$

$$D_2 = D_1 \sqrt{2}$$

9.25 A (materials with a relatively large bulk modulus). Usually, you would not want the volume of the building to change much when a force acts on it.

Estimation/Numerical Analysis

9.29 10×10^9 N/m² (approximately the same as for oak wood).

9.35

Strain (%)	Stress (10^9 N/m²)
0.0	0
0.1	125
0.2	250
0.3	230
0.4	230
0.5	235
0.6	240
0.7	250
0.8	260
0.9	270
1.0	300
1.5	325

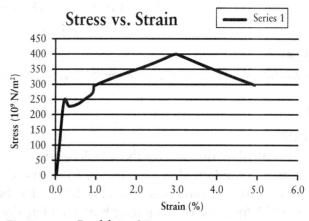

Figure 9-1 Problem 35

Strain (%)	Stress (10^9 N/m²)
2.0	350
2.5	375
3.0	400
3.5	375
4.0	350
4.5	325
5.0	300

Part a) The yield strength is the tensile stress at which the material becomes permanently deformed: 250×10^9 N/m².

Part b) The ultimate strength is the maximum tensile stress the material can withstand: 400×10^9 N/m².

Part c) The Young's modulus is the slope of the initial linear region: $125{,}000 \times 10^9$ N/m².

Get Help: P'Cast 9.6 – Human ACL I: Maximum Force
P'Cast 9.7 – Human ACL II: Breaking Strain

Problems

9.39

SET UP

An aluminum bar has a cross section of 1 cm² and an initial length of 88 cm. An applied force stretches the bar to a new length of 100 cm. We can calculate the magnitude of the force from the Young's modulus of aluminum and the tensile strain in the bar from the definition. The Young's modulus of aluminum from Table 9-1 is 70×10^9 N/m².

SOLVE

$$\frac{F}{A} = Y\frac{\Delta L}{L_0}$$

$$F = YA\frac{\Delta L}{L_0} = \left(70 \times 10^9 \frac{\text{N}}{\text{m}^2}\right)(10^{-4}\text{ m}^2)\left(\frac{(100\text{ cm}) - (88\text{ cm})}{88\text{ cm}}\right) = \boxed{1 \times 10^6 \text{ N} = 1 \times 10^3 \text{ kN}}$$

$$\frac{\Delta L}{L_0} = \frac{(100\text{ cm}) - (88\text{ cm})}{88\text{ cm}} = \boxed{0.1 = 10\%}$$

REFLECT

We would expect the force required to stretch an aluminum bar to be large.

9.43

SET UP

We can calculate the ratio of the tensile strains of an aluminum bar and a steel bar from their respective Young's moduli because the force and cross-sectional area are constant in both cases. The Young's modulus of steel is 200×10^9 N/m²; the Young's modulus of aluminum is 70×10^9 N/m². The original length of each bar does not affect the strain since the strain is essentially the percent change in the object's length.

SOLVE

Part a)

$$\frac{F}{A} = Y\frac{\Delta L}{L_0}$$

$$\frac{\left(\frac{\Delta L}{L_0}\right)_{Al}}{\left(\frac{\Delta L}{L_0}\right)_{Steel}} = \frac{\left(\frac{F}{AY}\right)_{Al}}{\left(\frac{F}{AY}\right)_{Steel}} = \frac{Y_{Steel}}{Y_{Al}} = \frac{\left(200\frac{N}{m^2}\right)}{\left(70\frac{N}{m^2}\right)} = \frac{20}{7} = \boxed{2.86}$$

Part b) The strain does not depend on the relative lengths of the bars (but the relative amount of compression does).

REFLECT

It makes sense that aluminum, with a smaller Young's modulus, should experience a larger strain than steel for a given force.

9.45

SET UP

A steel piano wire has a diameter of 0.20 cm. The tensile strength, which is the maximum tensile stress the material can withstand, of steel is 5.0×10^8 N/m². We can calculate the magnitude of the tension required to break the cylindrical wire directly from the strength of steel.

SOLVE

$$\text{ultimate strength} = \frac{F_{max}}{A}$$

$$5.0 \times 10^8 \frac{N}{m^2} = \frac{F_{max}}{A} = \frac{F_{max}}{\pi r^2}$$

$$F_{max} = \left(5.0 \times 10^8 \frac{N}{m^2}\right)(\pi r^2) = \left(5.0 \times 10^8 \frac{N}{m^2}\right)(\pi)\left(\frac{0.20 \times 10^{-2} \text{ m}}{2}\right)^2 = \boxed{1600 \text{ N} = 1.6 \text{ kN}}$$

REFLECT

The original length of the wire is irrelevant in this problem.

9.51

SET UP

A rigid cube of volume V_0 is filled with water and then frozen solid. The water expands by 9%, which means the volume changes by $+0.09V_0$. We can calculate the pressure exerted on the sides of the cube from the bulk modulus. The bulk modulus of water is 2.2×10^9 N/m².

SOLVE

$$\Delta p = -B\frac{\Delta V}{V_0}$$

$$|\Delta p| = B\frac{\Delta V}{V_0}$$

$$\frac{F}{A} = B\frac{\Delta V}{V_0} = B\frac{0.09 V_0}{V_0} = \left(2.2 \times 10^9 \frac{\text{N}}{\text{m}^2}\right)(0.09) = \boxed{2 \times 10^8 \frac{\text{N}}{\text{m}^2}}$$

REFLECT

We ignored the negative sign in the bulk modulus equation because we were just interested in the magnitude of the pressure.

Get Help: P'Cast 9.2 – A Submerged Cannonball
P'Cast 9.3 – Bubbles Rising

9.55

SET UP

A force of 200,000 N is acting tangentially to the top face of a brass plate. The top face is 2 cm by 20 cm. We can calculate the shear strain from the shear stress and the shear modulus. The shear modulus of brass from Table 9-1 is 40×10^9 N/m². The shear strain is equal to the tangent of φ.

Figure 9-2 Problem 55

SOLVE

$$\frac{F_{\parallel}}{A} = S\frac{x}{h}$$

$$\frac{x}{h} = \frac{F_{\parallel}}{AS} = \frac{200{,}000 \text{ N}}{(0.02 \text{ m})(0.20 \text{ m})\left(40 \times 10^9 \frac{\text{N}}{\text{m}^2}\right)} = \boxed{0.001}$$

$$\tan(\varphi) = \frac{x}{h}$$

$$\varphi = \tan^{-1}\left(\frac{x}{h}\right) = \tan^{-1} 0.00125 = \boxed{0.07°}$$

REFLECT

The top face will hang over the bottom face by about 100 microns.

Get Help: P'Cast 9.5 – Earthquake Damage

9.59

SET UP

In parts of the cardiovascular system, the cells lining the walls of the blood vessels experience a shear stress of 20 dyne/cm². We are told the shear strain is about 0.008. We can find the shear modulus directly from these data. We will need the conversion factor from dynes to newtons: $1 \text{ N} = 10^5$ dyne.

SOLVE

$$\frac{F_\parallel}{A} = S\frac{x}{h}$$

$$S = \frac{\left(\frac{F_\parallel}{A}\right)}{\left(\frac{x}{h}\right)} = \frac{\left(20\frac{\text{dyne}}{\text{cm}^2}\right)}{0.008} \times \frac{1 \text{ N}}{10^5 \text{ dyne}} \times \frac{10^4 \text{ cm}^2}{1 \text{ m}^2} = \boxed{3 \times 10^2 \frac{\text{N}}{\text{m}^2}}$$

REFLECT

We expect cells to have a much smaller shear modulus than, say, metal.

Get Help: P'Cast 9.5 – Earthquake Damage

9.61

SET UP

A given rope can support 10,000 N/m² before breaking. A knot is tied in it, which decreases the ultimate breaking strength by a factor of 2. A worker wants to use this knotted rope to hang a load of 1000 N/m² but needs to make sure it meets the safety standards set by his company. The company uses a safety factor of 10 for all of its ropes, which means the ultimate breaking strength of the rope must be at least 10 times larger than the load it is supporting. To determine the safety factor for using this rope to hold this load, we need to compare the maximum force the rope can support to the force of the load. If this is more than 10, then the worker can safely use the rope; if not, he'll need to find a new, stronger rope.

SOLVE

Part a)

The overestimated ultimate breaking strength of the knotted rope is

$$\frac{\left(10,000\frac{\text{N}}{\text{m}^2}\right)}{2} = 5000\frac{\text{N}}{\text{m}^2}$$

The safety factor in this case is

$$\frac{\left(5000\frac{\text{N}}{\text{m}^2}\right)}{\left(1000\frac{\text{N}}{\text{m}^2}\right)} = \boxed{5}$$

Part b) $\boxed{\text{No}}$, the rigging company cannot use this rope because the safety factor is less than 10.

Chapter 9 Elastic Properties of Matter: Stress and Strain

REFLECT
Since the cross section of the rope remains constant, the ratio between the ultimate breaking strength and the applied stress is equal to the ratio between the maximum force and the applied force.

Get Help: P'Cast 9.6 – Human ACL I: Maximum Force
P'Cast 9.7 – Human ACL II: Breaking Strain

General Problems

9.65

SET UP

Strands of caterpillar silk are typically 2.0 μm in diameter and have a Young's modulus of 4.0×10^9 N/m². We are asked to consider a climbing rope made of silk strands that is initially 9.0 m long and only stretches a distance of 1.00 cm when supporting the weight of two 85-kg climbers. We can calculate the cross-sectional area of this rope from the weight, the Young's modulus, and the strain. The ratio of this cross-sectional area to the cross-sectional area of one strand will give us the number of strands in the rope. Assuming the strands are distributed in a cylinder, we can use the area of a circle to calculate the diameter of the rope.

SOLVE
Part a)

$$\frac{F}{A} = Y\frac{\Delta L}{L_0}$$

$$A = \frac{FL_0}{Y\Delta L} = \frac{2(85 \text{ kg})\left(9.80 \frac{\text{m}}{\text{s}^2}\right)(9.0 \text{ m})}{\left(4.0 \times 10^9 \frac{\text{N}}{\text{m}^2}\right)(0.0100 \text{ m})} = 0.000375 \text{ m}^2$$

$$N_{\text{strands}} = \frac{A_{\text{total}}}{A_{1 \text{ strand}}} = \frac{0.000375 \text{ m}^2}{\pi\left(\frac{2.0 \times 10^{-6} \text{ m}}{2}\right)^2} = \boxed{1.2 \times 10^8 \text{ strands}}$$

Part b)

$$A_{\text{total}} = \pi\left(\frac{D}{2}\right)^2$$

$$D = \sqrt{\frac{4A_{\text{total}}}{\pi}} = \sqrt{\frac{4(0.000375 \text{ m}^2)}{\pi}} = \boxed{0.022 \text{ m} = 2.2 \text{ cm}}$$

Yes, this a reasonably-sized rope for mountain climbers to carry.

REFLECT
Climbing ropes made out of nylon have a comparable Young's modulus, 3.0×10^9 N/m².

9.69

SET UP

A weightlifter lifts a mass of 263.5 kg above his head. All of this weight is supported by the lifter's two tibias. The average length of a tibia is 385 mm and its diameter is 3.0 cm. The Young's modulus for bone is 2.0×10^{10} N/m². We can use these data to calculate the distance his tibia is compressed upon lifting the weight. In order to determine if that distance is significant, we need to compare it to the initial length of the tibia. Throughout the calculation, we will assume that the given length of the tibia (385 mm) includes compression due to his weight.

SOLVE

Part a)

$$\frac{F}{A} = Y\frac{\Delta L}{L_0}$$

$$\Delta L = \frac{FL_0}{AY} = \frac{(263.5 \text{ kg})\left(9.80\frac{\text{m}}{\text{s}^2}\right)(385 \text{ mm})}{2(\pi)\left(\frac{0.03 \text{ m}}{2}\right)^2\left(2.0 \times 10^{10}\frac{\text{N}}{\text{m}^2}\right)} = \boxed{0.035 \text{ mm}}$$

Part b) No, a change of 0.035 mm is not a significant compression.

Part c) No, we do not need to include his weight because we are interested in the compression due to extra weight. His tibia is initially compressed slightly due to his weight; we've assumed the value of 385 mm includes that initial compression.

REFLECT

If the value of 385 mm did *not* include the initial compression of his weight, his weight would compress the tibia by 0.02 mm, which is also negligible.

9.73

SET UP

The spherical bubbles near the surface of a glass of water have a diameter $D_0 = 2.5$ mm. This glass of water is taken to Golden, CO. The atmospheric pressure at sea level is 1.01×10^5 N/m², and the atmospheric pressure at Golden, CO, is 8.22×10^4 N/m². We can relate the change in the diameter and, thus, the volume to the change in atmospheric pressure through the bulk modulus.

SOLVE

$$\Delta p = -B\frac{\Delta V}{V_0}$$

$$\Delta p = -B\frac{\left(\frac{4}{3}\pi R^3\right) - \left(\frac{4}{3}\pi R_0^3\right)}{\left(\frac{4}{3}\pi R_0^3\right)} = -B\left(\frac{R^3}{R_0^3} - 1\right)$$

100 Chapter 9 Elastic Properties of Matter: Stress and Strain

$$R = R_0 \sqrt[3]{1 - \frac{\Delta p}{B}}$$

$$D_f = D_0 \sqrt[3]{1 - \frac{\Delta p}{B}} = (2.5 \text{ mm}) \sqrt[3]{1 - \frac{\left(\left(8.22 \times 10^4 \frac{N}{m^2}\right) - \left(1.01 \times 10^5 \frac{N}{m^2}\right)\right)}{\left(1.41 \times 10^5 \frac{N}{m^2}\right)}} = \boxed{2.6 \text{ mm}}$$

REFLECT
The pressure decreases by about 19%, whereas the diameter of the bubble increases by about 4%.

Get Help: P'Cast 9.5 – Earthquake Damage

Chapter 10
Gravitation

Conceptual Questions

10.1   Newton reasoned that the Moon, as it orbits, falls back toward the center of Earth due to the pull of gravity. At the same time that it moves tangent to the orbital path, it also falls back toward Earth. The combined motion leads to the familiar circular path. An apple falls straight toward Earth without the tangential motion. The vertical motion is the same for both the apple and the Moon.

   Get Help: Picture It – Gravitational Force

10.3   When the mass of one object is doubled, the force between two objects doubles. If both masses are halved, the force is one-fourth the original value.

   Get Help: Picture It – Gravitational Force

10.13 By Kepler's law, the faster an object is moving as it orbits the Sun, the closer it is to the Sun. Thus, Earth is closer to the Sun when the Northern Hemisphere is in winter.

10.15 Cells are probably the least affected by gravity (owing to strong electromagnetic forces at that level), but tissues and organs certainly depend on the constant supply of blood. This supply is intrinsically tied to the gravimetric forces on fluids in our body. Our vertebrae, for example, are continually compressed when we live on Earth. Our feet become flatter due to our continuous weight pushing down on them; our muscles sag over time, and so on.

Multiple-Choice Questions

10.21 **A** (she weighs more at the bottom). The value of g decreases with increasing altitude because the distance between Earth and the mountain climber is increasing.

   Get Help: Picture It – Gravitational Force
   P'Cast 10.1 – You and Your Backpack
   P'Cast 10.2 – Earth's Gravitational Force on You

10.23 **B** ($\sqrt{2}v$). We can calculate the speed of the satellite in the initial orbit using Newton's second law: $v = \sqrt{\dfrac{Gm_{\text{Earth}}}{r}}$. Comparing this to the escape speed formula, we see that the escape speed must be $\sqrt{2}v$.

   Get Help: P'Cast 10.6 – Escaping from a Martian Moon

10.25 **A** (Satellite A has more kinetic energy, less potential energy, and less mechanical energy (potential energy plus kinetic energy) than satellite B). The kinetic energy of an orbiting satellite is inversely proportional to the distance. Less potential energy in this case means "more negative."

Estimation/Numerical Analysis

10.27 Assuming it is possible to measure a force of about 10^{-7} N, the weights should have about 3 cm center-to-center separation. We can assume that it is possible to measure a force of $F = 10^{-7}$ N.

$$F = \frac{Gm_1 m_2}{r^2}$$

$$r = \sqrt{\frac{Gm_1 m_2}{F}} = \sqrt{\frac{\left(6.67 \times 10^{-11} \frac{\text{N} \cdot \text{m}^2}{\text{kg}^2}\right)(1 \text{ kg})^2}{(1 \times 10^{-7} \text{ N})}} \approx 3 \text{ cm}$$

Get Help: Picture It – Gravitational Force
P'Cast 10.1 – You and Your Backpack
P'Cast 10.2 – Earth's Gravitational Force on You

10.31 The force between Earth and Jupiter is 1/18,000 of the force between Earth and the Sun. The impact would be small.

Get Help: Picture It – Gravitational Force
P'Cast 10.1 – You and Your Backpack
P'Cast 10.2 – Earth's Gravitational Force on You

Problems

10.35

SET UP

A baseball ($m_{\text{ball}} = 0.150$ kg) is 100 m from a bat ($m_{\text{bat}} = 0.935$ kg). We can calculate the magnitude of the gravitational force between the two from Newton's universal law of gravitation.

SOLVE

$$F_{\text{ball on bat}} = \frac{Gm_{\text{ball}} m_{\text{bat}}}{r^2} = \frac{\left(6.67 \times 10^{-11} \frac{\text{N} \cdot \text{m}^2}{\text{kg}^2}\right)(0.150 \text{ kg})(0.935 \text{ kg})}{(100 \text{ m})^2} = \boxed{9.35 \times 10^{-16} \text{ N}}$$

REFLECT

As expected, this force is negligible.

Get Help: Picture It – Gravitational Force
P'Cast 10.1 – You and Your Backpack
P'Cast 10.2 – Earth's Gravitational Force on You

10.39

SET UP

We can use Newton's universal law of gravitation to calculate the net force acting on the Moon during an eclipse. The net force is equal to the vector sum of the gravitational force due to the Sun and the gravitational force due to Earth. In our coordinate system, we'll assume that all three satellites lie along the x-axis and that the Sun is to the left of the Moon (toward $-x$).

SOLVE

$$\sum F_{\text{ext},x} = F_{\text{Sun on Moon},x} + F_{\text{Earth on Moon},x} = -\frac{Gm_{\text{Sun}}m_{\text{Moon}}}{r_{\text{Sun to Moon}}^2} + \frac{Gm_{\text{Earth}}m_{\text{Moon}}}{r_{\text{Earth to Moon}}^2}$$

$$= Gm_{\text{Moon}}\left(\frac{m_{\text{Earth}}}{r_{\text{Earth to Moon}}^2} - \frac{m_{\text{Sun}}}{(r_{\text{Earth to Sun}} - r_{\text{Earth to Moon}})^2}\right)$$

$$= \left(6.67 \times 10^{-11}\frac{\text{N} \cdot \text{m}^2}{\text{kg}^2}\right)(7.35 \times 10^{22}\text{ kg})$$

$$\times \left(\frac{5.98 \times 10^{24}\text{ kg}}{3.84 \times 10^8\text{ m}^2} - \frac{1.99 \times 10^{30}\text{ kg}}{((1.50 \times 10^{11}\text{ m}) - (3.84 \times 10^8\text{ m}))^2}\right)$$

$$= -2.37 \times 10^{20}\text{ N}$$

The net force acting on the Moon has a magnitude of $\boxed{2.37 \times 10^{20}\text{ N and points toward the Sun}}$.

REFLECT

The gravitational force on the Moon due to the Sun is still larger than the force due to Earth even though Earth is much closer.

Get Help: Picture It – Gravitational Force
P'Cast 10.1 – You and Your Backpack
P'Cast 10.2 – Earth's Gravitational Force on You

10.41

SET UP

We are interested in the point where the net gravitational force due to Earth ($m_{\text{Earth}} = 5.98 \times 10^{24}$ kg) and the Moon ($m_{\text{Moon}} = 7.35 \times 10^{22}$ kg) on a space probe (mass m_{probe}) is equal to zero; the point where the net force is equal to zero must be between Earth and the Moon. The Moon is 3.84×10^8 m from Earth. We will let x be the distance from Earth to the probe and y be the distance from the Moon to the probe. We'll assume that Earth is to the left of the probe and the Moon is to the right.

SOLVE

$$\frac{Gm_{\text{Earth}}m_{\text{probe}}}{x^2} = \frac{Gm_{\text{Moon}}m_{\text{probe}}}{y^2}$$

$$\frac{x^2}{y^2} = \frac{m_{\text{Earth}}}{m_{\text{Moon}}}$$

$$x = y\sqrt{\frac{m_{\text{Earth}}}{m_{\text{Moon}}}}$$

$$x + y = 3.84 \times 10^8 \text{ m}$$

$$y\sqrt{\frac{m_{\text{Earth}}}{m_{\text{Moon}}}} + y = 3.84 \times 10^8 \text{ m}$$

$$y = \frac{3.84 \times 10^8 \text{ m}}{1 + \sqrt{\frac{m_{\text{Earth}}}{m_{\text{Moon}}}}} = \frac{3.84 \times 10^8 \text{ m}}{1 + \sqrt{\frac{5.98 \times 10^{24} \text{ kg}}{7.35 \times 10^{22} \text{ kg}}}} = 3.83 \times 10^7 \text{ m}$$

$$x = (3.84 \times 10^8 \text{ m}) - y = 3.46 \times 10^8 \text{ m}$$

The point where the net gravitational force due to Earth and the Moon is equal to zero is $\boxed{3.46 \times 10^8 \text{ m to the right of Earth}}$ or $\boxed{3.83 \times 10^7 \text{ m to the left of the Moon}}$.

REFLECT

The exact mass of the space probe is not necessary to solve this problem. It makes sense that the point where the net force is zero should be closer to the Moon than Earth since it is less massive.

Get Help: Picture It – Gravitational Force
P'Cast 10.1 – You and Your Backpack
P'Cast 10.2 – Earth's Gravitational Force on You

10.45

SET UP

A 10-kg object moves from a point 6000 m above sea level to a point 1000 m above sea level. The work done by gravity is equal to the negative of the change in the gravitational potential energy of the object. Because the gravitational force from a spherically symmetric object is the same as if all its mass were concentrated at its center, we need to include the radius of Earth when determining the initial and final positions.

SOLVE

$$W_{\text{gravity on object}} = -\Delta U_{\text{grav}} = -\left(\frac{-Gm_{\text{Earth}}m_{\text{object}}}{r_f} - \left(\frac{-Gm_{\text{Earth}}m_{\text{object}}}{r_i}\right)\right)$$

$$= Gm_{\text{Earth}}m_{\text{object}}\left(\frac{1}{R_{\text{Earth}} + (1000 \text{ m})} - \frac{1}{R_{\text{Earth}} + (6000 \text{ m})}\right)$$

$$= \left(6.67 \times 10^{-11} \frac{\text{N} \cdot \text{m}^2}{\text{kg}^2}\right)(5.98 \times 10^{24} \text{ kg})(10 \text{ kg})$$

$$\left(\frac{1}{(6.38 \times 10^6 \text{ m}) + (1000 \text{ m})} - \frac{1}{(6.38 \times 10^6 \text{ m}) + (6000 \text{ m})}\right)$$

$$= \boxed{489{,}400 \text{ J} = 5 \times 10^2 \text{ kJ}}$$

REFLECT

The force of gravity and the displacement are in the same direction, so we expect the work done by gravity to be positive. If we used $U_{\text{grav}} = mgy$ instead, the work done by gravity would be about the same.

10.47

SET UP

We can use conservation of energy to calculate the impact speed of a 100-kg asteroid that is initially moving at a speed of 200 m/s at a distance of 1000 km from the Moon's surface. When calculating the initial and final gravitational potential energies of the asteroid, we need to include the radius of the Moon ($r_{Moon} = 1.737 \times 10^6$ m). The force of the Moon on the asteroid is nonconservative and the only force doing work to stop the asteroid. Therefore, the work done by the Moon on the asteroid is equal to the change in the asteroid's mechanical energy. We will assume that the asteroid does not move very deeply into the surface during the impact, so $\Delta U_{grav} \approx 0$.

SOLVE
Part a)

$$U_{grav,i} + K_i = U_{grav,f} + K_f$$

$$-\frac{Gm_{Moon}m_{asteroid}}{r_i} + \frac{1}{2}m_{asteroid}v_i^2 = -\frac{Gm_{Moon}m_{asteroid}}{r_f} + \frac{1}{2}m_{asteroid}v_f^2$$

$$v_f = \sqrt{-\frac{2Gm_{Moon}}{r_i} + \frac{2Gm_{Moon}}{r_f} + v_i^2} = \sqrt{-2Gm_{Moon}\left(\frac{1}{(r_{Moon} + (10^6 \text{ m}))} - \frac{1}{r_{Moon}}\right) + v_i^2}$$

$$= \sqrt{-2\left(6.67 \times 10^{-11} \frac{\text{N} \cdot \text{m}^2}{\text{kg}^2}\right)(7.35 \times 10^{22} \text{ kg})\left(\frac{1}{((1.737 \times 10^6 \text{ m}) + (10^6 \text{ m}))} - \frac{1}{(1.737 \times 10^6 \text{ m})}\right) + \left(200\frac{\text{m}}{\text{s}}\right)^2}$$

$$= \boxed{1450\frac{\text{m}}{\text{s}} = 1 \times 10^3 \frac{\text{m}}{\text{s}}}$$

Part b)

$$W_{nonconservative} = \Delta K + \Delta U_{grav} = \frac{1}{2}m_{asteroid}(v_f^2 - v_i^2) + 0 = \frac{1}{2}m_{asteroid}(0 - v_i^2)$$

$$= -\frac{1}{2}(100 \text{ kg})\left(1450\frac{\text{m}}{\text{s}}\right)^2 = -1.05 \times 10^8 \text{ J}$$

The Moon does $\boxed{1 \times 10^8 \text{ J}}$ of work.

REFLECT

We would expect the asteroid to heat up and/or break apart upon impact as well as embed itself into the Moon's surface, so our calculation gives an upper bound to the work done by the Moon.

10.53

SET UP

A satellite orbits Earth once every 86.5 min. Assuming its orbit is circular, we can use Kepler's law of periods to calculate the radius of the orbit, and then the circumference of the satellite's orbit, from the period.

SOLVE

$$T^2 = \frac{4\pi^2 r^3}{Gm_{\text{Earth}}}$$

$$r = \sqrt[3]{\frac{Gm_{\text{Earth}}T^2}{4\pi^2}} = \sqrt[3]{\frac{\left(6.67 \times 10^{-11}\frac{\text{N}\cdot\text{m}^2}{\text{kg}^2}\right)(5.98 \times 10^{24}\text{ kg})\left(86.5 \text{ min} \times \frac{60 \text{ s}}{1 \text{ min}}\right)^2}{4\pi^2}}$$

$$= 6.48 \times 10^6 \text{ m}$$

$$C = 2\pi r = 2\pi(6.48 \times 10^6 \text{ m}) = \boxed{4.07 \times 10^7 \text{ m}}$$

REFLECT
Be sure to use a consistent set of units when performing your calculations.

10.57

SET UP
The Moon orbits Earth in a nearly circular orbit once every 27.32 days. We can use Kepler's law of periods to help calculate the distance d from the surface of the Moon to the surface of Earth. The radius in Kepler's law is the center-to-center distance between the Moon and Earth, so we will need to subtract out the radius of Earth and the radius of the Moon in order to find d.

SOLVE
Converting the period into seconds:

$$27.32 \text{ d} \times \frac{24 \text{ h}}{1 \text{ d}} \times \frac{3600 \text{ s}}{1 \text{ h}} = 2.360 \times 10^6 \text{ s}$$

Calculating the distance:

$$T^2 = \frac{4\pi^2 r^3}{Gm_{\text{Earth}}} = \frac{4\pi^2 (R_{\text{Earth}} + R_{\text{Moon}} + d)^3}{Gm_{\text{Earth}}}$$

$$d = \sqrt[3]{\frac{Gm_{\text{Earth}}T^2}{4\pi^2}} - R_{\text{Earth}} - R_{\text{Moon}}$$

$$= \sqrt[3]{\frac{\left(6.67 \times 10^{-11}\frac{\text{N}\cdot\text{m}^2}{\text{kg}^2}\right)(5.98 \times 10^{24}\text{ kg})(2.360 \times 10^6 \text{ s})^2}{4\pi^2}}$$

$$- (6.38 \times 10^6 \text{ m}) - (1.73 \times 10^6 \text{ m})$$

$$= \boxed{3.75 \times 10^8 \text{ m}}$$

REFLECT
The mean distance from the center of Earth to the center of the Moon is 3.84×10^8 m. This is larger than the distance between their surfaces, which makes sense.

General Problems

10.61

SET UP

A rocket is launched into a low-Earth orbit ($r \approx R_{Earth}$) near the equator. We can use conservation of energy to calculate the necessary launch speed of the rocket v_{launch}. A rocket launched due east is launched with the rotation of Earth, while a rocket launched due west is launched against the rotation of Earth, so the rotation speed of Earth will either be added or subtracted to the launch speed v_{launch} of the rocket.

SOLVE

Tangential speed of Earth's surface:

$$v_{Earth} = \frac{2\pi R_{Earth}}{T} = \frac{2\pi(6.38 \times 10^6 \text{ m})}{\left(24 \text{ h} \times \frac{3600 \text{ s}}{1 \text{ h}}\right)} = 464 \frac{\text{m}}{\text{s}}$$

Part a)

$$K_i + U_{grav,i} = K_f + U_{grav,f}$$

$$\frac{1}{2}m_{rocket}\left(v_{launch} + \left(464\frac{\text{m}}{\text{s}}\right)\right)^2 - \frac{Gm_{Earth}m_{rocket}}{R_{Earth}} = 0$$

$$v_{launch} = \sqrt{\frac{2Gm_{Earth}}{R_{Earth}}} - \left(464\frac{\text{m}}{\text{s}}\right) = \sqrt{\frac{2\left(6.67 \times 10^{-11}\frac{\text{N} \cdot \text{m}^2}{\text{kg}^2}\right)(5.98 \times 10^{24} \text{ kg})}{(6.38 \times 10^6 \text{ m})}} - \left(464\frac{\text{m}}{\text{s}}\right)$$

$$= \boxed{1.07 \times 10^4 \frac{\text{m}}{\text{s}} = 10.7 \frac{\text{km}}{\text{s}}}$$

Part b)

$$K_i + U_{grav,i} = K_f + U_{grav,f}$$

$$\frac{1}{2}m_{rocket}\left(v_{launch} - \left(464\frac{\text{m}}{\text{s}}\right)\right)^2 - \frac{Gm_{Earth}m_{rocket}}{R_{Earth}} = 0$$

$$v_{launch} = \sqrt{\frac{2Gm_{Earth}}{R_{Earth}}} + \left(464\frac{\text{m}}{\text{s}}\right) = \sqrt{\frac{2\left(6.67 \times 10^{-11}\frac{\text{N} \cdot \text{m}^2}{\text{kg}^2}\right)(5.98 \times 10^{24} \text{ kg})}{(6.38 \times 10^6 \text{ m})}} + \left(464\frac{\text{m}}{\text{s}}\right)$$

$$= \boxed{1.16 \times 10^4 \frac{\text{m}}{\text{s}} = 11.6 \frac{\text{km}}{\text{s}}}$$

REFLECT

It makes sense that a rocket needs to launch with a smaller speed if it launches with the rotation of Earth.

10.69

SET UP

Two objects ($m_{10} = 10.0$ kg and $m_3 = 3.00$ kg) are separated by 40.0 cm. A third object ($m_1 = 1.00$ kg) is placed at a location along the line connecting them such that the net force acting on m_1 is zero. By considering the force vectors, this location must be between the two original objects. We will define x as the distance between m_{10} and m_1 and y as the distance between m_3 and m_1. Setting the magnitudes of the gravitational forces equal and using the fact that $x + y = 40.0$ cm, we can solve for x.

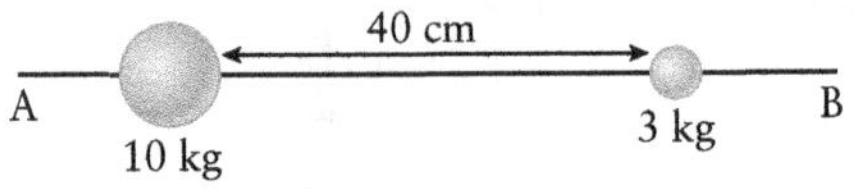

Figure 10-1 Problem 69

SOLVE

$$\frac{Gm_{10}m_1}{x^2} = \frac{Gm_3m_1}{y^2}$$

$$x^2 = \frac{m_{10}y^2}{m_3}$$

$$x = y\sqrt{\frac{m_{10}}{m_3}}$$

But $x + y = 40$ cm:

$$x + y = y\sqrt{\frac{m_{10}}{m_3}} + y = y\sqrt{\frac{10.0 \text{ kg}}{3.00 \text{ kg}}} + y = 2.826y = 40.0 \text{ cm}$$

$$y = \frac{40.0 \text{ cm}}{2.826} = 14.2 \text{ cm}$$

$$x = (40.0 \text{ cm}) - y = (40.0 \text{ cm}) - (14.2 \text{ cm}) = \boxed{25.8 \text{ cm}}$$

REFLECT

It makes sense that the 1-kg object needs to be closer to the less massive object if the net force acting on it is zero.

Get Help: Picture It – Gravitational Force
P'Cast 10.1 – You and Your Backpack
P'Cast 10.2 – Earth's Gravitational Force on You

Chapter 11
Fluids

Conceptual Questions

11.3 No. The pressure on the water is determined by its depth, gravity, and the pressure pushing down on the surface (air pressure). The ability of the dam to withstand that pressure does, of course, depend on its shape.

11.9 It is easier to float in the Great Salt Lake because salt water is denser than freshwater.

      Get Help: Picture It – Buoyancy Force
          Interactive Exercise – Bobber
          Interactive Exercise – Floating Cylinders

11.15 The wind flow over the roof is very fast, while the airflow inside the house is not; this lowers the exterior pressure. If the roof is not made to hold down against a large part of an atmosphere of pressure pushing it up, it will fly off.

Multiple-Choice Questions

11.19 D (be increased, but not necessarily doubled). Because the gauge pressure is the pressure above 1 atm, doubling the gauge pressure will not necessarily double the absolute pressure.

11.23 A (one-quarter the speed in the 0.5-cm pipe).

$$A_1 v_1 = A_2 v_2$$

$$v_2 = v_1 \frac{A_1}{A_2} = v_1 \left(\frac{\pi (0.25 \text{ cm})^2}{\pi (0.5 \text{ cm})^2} \right) = \boxed{\frac{1}{4} v_1}$$

      Get Help: P'Cast 11.11 – Flow in a Constriction

11.27 A (Box A).

$$\sum F_y = F_b - w_{object} = \rho_{water} V_{displaced} g - \rho_{object} V_{object} g = m a_y = \rho_{object} V_{object} a_y$$

Since $V_{object} = V_{displaced}$, we can solve for a_y:

$$a_y = \frac{\rho_{water} V_{object} g - \rho_{object} V_{object} g}{\rho_{object} V_{object}} = g \left(\frac{\rho_{water}}{\rho_{object}} - 1 \right)$$

The object with the smaller density will experience a larger acceleration upward.

      Get Help: Picture It – Buoyancy Force
          Interactive Exercise – Bobber
          Interactive Exercise – Floating Cylinders

Estimation/Numerical Analysis

11.29 A wind that is 33 m/s (barely hurricane level) passing over the top of a 1-m² prone body resting on top of still air results in a lift of approximately 650 N.

11.31 The required depth for mercury is, of course, 760 mm; for water it is 13.6 times greater, or 10.3 m; and for seawater it is 10.0 m.

Get Help: P'Cast 11.4 – Air Pressure versus Water Pressure

Problems

11.37

SET UP

A cylinder made of an unknown material is 20 cm long, 1 cm in radius, and has a mass of 37 g. We can calculate the average density of the material by dividing the cylinder's mass by its volume. The volume of a cylinder is the cross-sectional area multiplied by the length. We can compare our answer to part (a) to the densities in Table 11-1 in order to figure out the material from which the cylinder is made. The specific gravity of a material is equal to its density divided by the density of water, $\rho_{water} = 1.000 \times 10^3$ kg/m³.

SOLVE

$$\rho = \frac{m}{V} = \frac{m}{\pi R^2 L} = \frac{(0.037 \text{ kg})}{\pi (0.01 \text{ m})^2 (0.2 \text{ m})} = \boxed{6 \times 10^2 \frac{\text{kg}}{\text{m}^3}}$$

Part b)
The cylinder is most likely made of fresh wood.

Part c)

$$SG = \frac{\rho}{\rho_{water}} = \frac{6 \times 10^2 \frac{\text{kg}}{\text{m}^3}}{1.000 \times 10^3 \frac{\text{kg}}{\text{m}^3}} = \boxed{0.6}$$

REFLECT

A "uniform" object has a constant density throughout it. If an object is not uniform, we can calculate the average density of the object as we did above.

Get Help: P'Cast 11.1 – Density of Chicken

11.41

SET UP

A neutron star has the same density as a neutron. We can calculate the radius of a neutron star that has the same mass as our Sun ($m_{star} = 1.99 \times 10^{30}$ kg) by setting the ratio of its mass to volume equal to the ratio of a neutron's mass to volume. We will assume both the star and the neutron are spheres. The mass of a neutron is $m_{neutron} = 1.7 \times 10^{-27}$ kg, and its approximate radius is 1.2×10^{-15} m.

SOLVE

$$\rho_{\text{neutron}} = \rho_{\text{star}}$$

$$\frac{m_{\text{neutron}}}{\left(\frac{4}{3}\pi R^3_{\text{neutron}}\right)} = \frac{m_{\text{star}}}{\left(\frac{4}{3}\pi R^3_{\text{star}}\right)}$$

$$R_{\text{star}} = R_{\text{neutron}}\sqrt[3]{\frac{m_{\text{star}}}{m_{\text{neutron}}}} = (1.2 \times 10^{-15} \text{ m})\sqrt[3]{\frac{1.99 \times 10^{30} \text{ kg}}{1.7 \times 10^{-27} \text{ kg}}} = \boxed{13 \text{ km}}$$

REFLECT

A radius of 13 km seems reasonable for a "very small" star. Most astronomical distances are orders of magnitude larger. For instance, the radius of Earth is approximately 6400 km and the distance between Earth and the Moon is about 384,000 km.

Get Help: P'Cast 11.1 – Density of Chicken

11.45

SET UP

A force of 25 N is applied to the head of a nail that is 0.32 cm in diameter. The pointed end of the nail is 0.032 cm in diameter. The pressure on the head of the nail or the pointed end of the nail is equal to the applied force divided by the cross-sectional area at that location.

SOLVE
Part a)

$$p = \frac{F_\perp}{A} = \frac{F_\perp}{\pi\left(\frac{d}{2}\right)^2} = \frac{4F_\perp}{\pi d^2} = \frac{4(25 \text{ N})}{\pi(0.32 \times 10^{-2} \text{ m})^2} = \boxed{3.1 \times 10^6 \text{ Pa} = 3.1 \text{ MPa}}$$

Part b)

$$p = \frac{F_\perp}{A} = \frac{F_\perp}{\pi\left(\frac{d}{2}\right)^2} = \frac{4F_\perp}{\pi d^2} = \frac{4(25 \text{ N})}{\pi(0.032 \times 10^{-2} \text{ m})^2} = \boxed{3.1 \times 10^8 \text{ Pa} = 310 \text{ MPa}}$$

REFLECT

Since the pressure is inversely proportional to the area, a decrease of a factor of 10 in the radius corresponds to an increase by a factor of 100 in the pressure!

11.49

SET UP

The difference in blood pressure between the top of the head and the bottom of the feet of a 1.75-m-tall person is equal to product of the density of blood ($\rho_{\text{blood}} = 1.06 \times 10^3$ kg/m^3), g, and the height. The conversion between pascals and mmHg is 1.01×10^5 Pa = 760 mmHg.

Chapter 11 Fluids

SOLVE

$$p = p_0 + \rho g d$$

$$\Delta p = p_{head} - p_{feet} = (p_0 + \rho_{blood} g d_{head}) - (p_0 + \rho_{blood} g d_{feet}) = \rho_{blood} g h - \rho_{blood} g(0)$$

$$\Delta p = \rho_{blood} g h = \left(1.06 \times 10^3 \frac{kg}{m^3}\right)\left(9.80 \frac{m}{s^2}\right)(1.75 \text{ m})$$

$$= 18179 \text{ Pa} \times \frac{760 \text{ mmHg}}{1.01 \times 10^5 \text{ Pa}} = \boxed{137 \text{ mmHg}}.$$

REFLECT

Since we are just looking for the difference in blood pressure between the top of the head and the bottom of the feet, we do not need to explicitly include the pressure of the atmosphere in our calculation.

Get Help: P'Cast 11.4 – Air Pressure versus Water Pressure

11.51

SET UP

Blood pressure is usually measured as a gauge pressure, which is the pressure above the atmospheric pressure. To convert a gauge pressure into an absolute pressure, we need to add the pressure of the atmosphere. In mmHg, the atmospheric pressure is 760 mmHg.

SOLVE

$$p = p_{gauge} + p_{atm} = (120 \text{ mmHg}) + (760 \text{ mmHg}) = \boxed{880 \text{ mmHg}}$$

REFLECT

Two common examples in which gauge pressures are used are for blood pressures and for car tires. Be sure to convert to absolute pressure by adding the atmospheric pressure before using the pressures in an equation.

11.57

SET UP

An airplane window has an area of 1000 cm². The pressure inside the airplane is 0.95 atm, while the pressure outside the airplane is 0.85 atm. The net force due to this pressure difference will point from the location of high pressure (inside the plane, in this case) toward the location of low pressure and have a magnitude equal to the pressure difference multiplied by the cross-sectional area of the window.

SOLVE

$$F_{net} = (p_2 - p_1)A$$

$$F_{net} = (0.95 \text{ atm} - 0.85 \text{ atm})\left(\frac{1.01 \times 10^5 \text{ Pa}}{1 \text{ atm}}\right)\left(1000 \text{ cm}^2 \times \frac{1 \text{ m}^2}{10^4 \text{ cm}^2}\right)$$

$$F_{net} = 1.0 \times 10^3 \text{ N} = 1.0 \text{ kN}$$

The net force on the airplane window has a magnitude of $\boxed{1.0 \text{ kN and points outward}}$.

REFLECT

The net force due to a pressure difference always points from high pressure to low pressure.

Get Help: P'Cast 11.6 – Submarine Hatch

11.59

SET UP

The hatch on the Mars lander is built and tested on Earth such that the internal pressure of the lander exactly balances the external pressure (that is, atmospheric pressure). The lander is then brought to Mars, where the external pressure is 650 Pa. The net force acting on the hatch is equal to the difference in the internal and external pressures multiplied by the cross-sectional area of the hatch. The hatch is round with a diameter of 0.500 m. Since the internal pressure of the lander is larger than the external pressure on Mars, the net force will point outward.

SOLVE

$$F_{net} = (p_2 - p_1)A$$

$$F_{net} = (\Delta p)A = ((1.01 \times 10^5 \text{ Pa}) - (650 \text{ Pa}))(\pi)\left(\frac{0.500 \text{ m}}{2}\right)^2$$

$$= \boxed{1.97 \times 10^4 \text{ N} = 2.0 \times 10^4 \text{ N}}$$

The force in pounds is:

$$1.97 \times 10^4 \text{ N} \times \frac{0.22 \text{ lb}}{1 \text{ N}} = 4.3 \times 10^3 \text{ lb}$$

This force points $\boxed{\text{outward}}$.

REFLECT

We can check our answer by looking at the orders of magnitude of each quantity:

$$F_{net} = (\Delta p)A \approx (10^5)\frac{(5 \times 10^{-1})^2}{4} = (10^5)\frac{(25)}{4}(10^{-2}) = (10^5)(10^1)(10^{-2}) = 10^4 \text{ N}$$

Get Help: P'Cast 11.6 – Submarine Hatch

11.63

SET UP

A hydraulic lift is made up of a small piston of area $A_1 = 0.0330 \text{ m}^2$ and a large piston of area $A_2 = 4.00 \text{ m}^2$. An applied force of $F_1 = 16.0 \text{ N}$ causes the small piston to move downward a distance Δy_1. Pascal's principle lets us calculate the force that the large piston provides. The volume of the fluid moved on the left side must equal the volume of the fluid moved on the right side. We can find Δy_2 in terms of Δy_1 by setting these volumes equal to one another. The work done in moving each piston is equal to the magnitude of the force applied to the piston multiplied by the distance through which the piston travels.

114 Chapter 11 Fluids

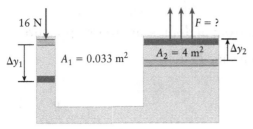

Figure 11-1 Problem 63

SOLVE
Part a)
$$p_1 = p_2$$
$$\frac{F_1}{A_1} = \frac{F_2}{A_2}$$
$$F_2 = F_1 \frac{A_2}{A_1} = (16 \text{ N})\frac{(4 \text{ m}^2)}{(0.033 \text{ m}^2)} = \boxed{1940 \text{ N} = 2 \times 10^3 \text{ N}}$$

Part b)
$$A_1 \Delta y_1 = A_2 \Delta y_2$$
$$\Delta y_2 = \frac{A_1}{A_2}\Delta y_1 = \left(\frac{0.033 \text{ m}^2}{4 \text{ m}^2}\right)\Delta y_1 = \boxed{0.00825 \Delta y_1 = 8 \times 10^{-3} \Delta y_1}$$

Part c)
$$W_1 = F_1 \Delta y_1 = (16 \text{ N})(0.20 \text{ m}) = \boxed{3.2 \text{ J}}$$
$$W_2 = F_2 \Delta y_2 = F_2(0.00825 \Delta y_1) = (1940 \text{ N})(0.00825)(0.20 \text{ m}) = \boxed{3.2 \text{ J}}$$

The work done in slowly pushing the small piston is $\boxed{\text{equal}}$ to the work done in raising the large piston.

REFLECT
Since the density and mass of the liquid inside the lift remain constant, the volume must also remain constant.

11.69

SET UP
A crown is weighed in air and in water. In air, the weight of the crown is measured as 5.15 N; we can calculate the mass of the crown from this measurement. In water, the apparent weight is 4.88 N. The apparent weight is equal to the weight of the crown in air minus the magnitude of the buoyant force, which is related to the volume of the crown. From both of these measurements, we can calculate the average density of the crown and compare it to the density of gold, which is 19.3 times the density of water. If the average density of the crown is close to the density of gold, then the crown is most likely made of gold.

SOLVE
$$w_{\text{crown}} = m_{\text{crown}} g$$

$$m_{crown} = \frac{w_{crown}}{g} = \frac{5.15 \text{ N}}{9.80\frac{\text{m}}{\text{s}^2}} = 0.526 \text{ kg}$$

$$w_{app} = w_{crown} - F_b = w_{crown} - \rho_{water} g V_{crown}$$

$$V_{crown} = \frac{w_{crown} - w_{app}}{\rho_{water} g} = \frac{(5.15 \text{ N}) - (4.88 \text{ N})}{\left(1000\frac{\text{kg}}{\text{m}^3}\right)\left(9.80\frac{\text{m}}{\text{s}^2}\right)} = 2.76 \times 10^{-5} \text{ m}^3$$

$$\rho_{crown} = \frac{m_{crown}}{V_{crown}} = \frac{0.526 \text{ kg}}{2.76 \times 10^{-5} \text{ m}^3} = 1.91 \times 10^4 \frac{\text{kg}}{\text{m}^3}$$

$$SG_{crown} = \frac{\rho_{crown}}{\rho_{water}} = \frac{1.91 \times 10^4 \frac{\text{kg}}{\text{m}^3}}{1000\frac{\text{kg}}{\text{m}^3}} = 19.1$$

This is close to the specific gravity of gold, so the crown is most likely made of gold.

REFLECT

The crown could still be fake, though. The density we calculated is the average density of the crown, which means it could be made up of a mixture of less dense and more dense materials.

Get Help: Picture It – Buoyancy Force
Interactive Exercise – Bobber
Interactive Exercise – Floating Cylinders
P'Cast 11.9 – Underwater Float

11.71

SET UP

A hose with a radius of 1 cm is connected to a faucet and used to fill a 5.0-L container in 45 s. The volume flow rate is equal to the total volume filled divided by the time it took to do so. The volume flow rate is also equal to the cross-sectional area of the hose multiplied by the speed of the water in the hose.

SOLVE

Part a)

$$Q = \frac{V}{t} = \frac{5.0 \text{ L}}{45 \text{ s}} \times \frac{1 \text{ m}^3}{1000 \text{ L}} = \boxed{1.1 \times 10^{-4} \frac{\text{m}^3}{\text{s}}}$$

Part b)

$$Q = Av = \pi R^2 v$$

$$v = \frac{Q}{\pi R^2} = \frac{1.1 \times 10^{-4} \frac{\text{m}^3}{\text{s}}}{\pi (1 \times 10^{-2} \text{ m}^2)^2} = \boxed{0.4 \frac{\text{m}}{\text{s}}}$$

116 Chapter 11 Fluids

REFLECT

The volume flow rate will be constant since the hose is a closed system.

Get Help: P'Cast 11.11 – Flow in a Constriction

11.75

SET UP

The inner diameter of a needle used for an injection is 0.114×10^{-3} m. An injection of 2.5×10^{-3} L was given in 0.65 s. We can equate the two expressions for the volume flow rate—namely, the total volume divided by the time and the cross-sectional area of the needle multiplied by the speed of the fluid—and solve for the speed of the fluid as it leaves the needle.

SOLVE

$$Q = \frac{V}{t} = Av$$

$$v = \frac{V}{At} = \frac{\left(2.5 \times 10^{-3} \text{ L} \times \frac{1 \text{ m}^3}{1000 \text{ L}}\right)}{\pi \left(\frac{0.114 \times 10^{-3} \text{ m}}{2}\right)^2 (0.65 \text{ s})} = \boxed{3.8 \times 10^2 \frac{\text{m}}{\text{s}}}$$

REFLECT

This large speed is a reason why most injections are given very slowly.

Get Help: P'Cast 11.11 – Flow in a Constriction

11.79

SET UP

At one point, the maximum wind speed during Hurricane Katrina was 240 km/h and the pressure in the eye was 0.877 atm. We can use the Bernoulli equation and the measured wind speed to calculate the theoretical pressure inside the eye. We will assume the pressure outside of the eye is 1 atm.

SOLVE

$$p_1 + \frac{1}{2}\rho v_1^2 + \rho g y_1 = p_2 + \frac{1}{2}\rho v_2^2 + \rho g y_2$$

$$p_{atm} + 0 + 0 = p_{eye} + \frac{1}{2}\rho v_{eye}^2 + 0$$

$$p_{eye} = p_{atm} - \frac{1}{2}\rho v_{eye}^2 = (1.01 \times 10^5 \text{ Pa}) - \frac{1}{2}\left(1.23 \frac{\text{kg}}{\text{m}^3}\right)\left(240 \frac{\text{km}}{\text{h}} \times \frac{1000 \text{ m}}{1 \text{ km}} \times \frac{1 \text{ h}}{3600 \text{ s}}\right)^2$$

$$= 98{,}267 \text{ Pa} \times \frac{1 \text{ atm}}{1.01 \times 10^5 \text{ Pa}} = \boxed{0.973 \text{ atm}}$$

This is larger than the measured value of 0.877 atm. The pressure outside of the eye is most likely not 1 atm.

REFLECT
The wind is made to go in a circle by the low pressure in the still center, so most of the pressure drop is due to a lower baseline pressure (that is, less than 1 atm), not the Bernoulli effect.

11.83

SET UP
Blood takes about 1.50 s to pass through a capillary ($L = 2.00 \times 10^{-3}$ m) in the human circulatory system. The capillary has a radius of $R = 2.50 \times 10^{-6}$ m. The pressure drop across the capillary is $\Delta p = 2.60 \times 10^3$ Pa. We can use Poiseuille's law in order to calculate the viscosity of the blood. The volumetric flow rate is equal to the volume of the blood in the capillary, which is essentially equal to the volume of the capillary, divided by the time it takes the blood to pass through the capillary. For simplicity, we'll assume the capillary is a cylinder.

SOLVE

$$Q = \frac{\pi R^4}{8\eta L}\Delta p$$

$$\eta = \frac{\pi R^4}{8LQ}\Delta p = \frac{\pi R^4}{8L\left(\frac{V}{t}\right)}\Delta p = \frac{\pi R^4 t}{8L(\pi R^2 L)}\Delta p = \frac{R^2 t}{8L^2}\Delta p = \frac{\left(\frac{D}{2}\right)^2 t}{8L^2}\Delta p$$

$$= \frac{D^2 t}{32L^2}\Delta p = \frac{(5.00 \times 10^{-6}\text{ m})^2(1.50\text{ s})}{32(2.00 \times 10^{-3}\text{ m})^2}(2.60 \times 10^3\text{ Pa}) = \boxed{7.62 \times 10^{-4}\frac{\text{N}\cdot\text{s}}{\text{m}^2}}$$

REFLECT
This is a little less than the value listed in Table 11-2 for the viscosity of blood at body temperature $\left(\sim 3 \times 10^{-4}\frac{\text{N}\cdot\text{s}}{\text{m}^2}\right)$.

General Problems

11.89

SET UP
A normal blood pressure is reported as 120/80, where the top number is the systolic pressure and the bottom number is the diastolic pressure. Both of these values are given in mmHg. We can rewrite these pressures in terms of other pressure units by applying the following conversion factors: 760 mmHg = 1.01×10^5 Pa = 1 atm = 14.7 psi. Blood pressure is an example of a gauge pressure, as 120 mmHg and 80 mmHg refer to the pressures *above* atmospheric pressure.

SOLVE
Part a)

Systolic:

$$120\text{ mmHg} \times \frac{1.01 \times 10^5\text{ Pa}}{760\text{ mmHg}} = 1.59 \times 10^4\text{ Pa}$$

Diastolic:

$$80 \text{ mmHg} \times \frac{1.01 \times 10^5 \text{ Pa}}{760 \text{ mmHg}} = 1.06 \times 10^4 \text{ Pa}$$

Blood pressure:

$$\boxed{\begin{array}{c} 1.59 \times 10^4 \text{ Pa} \\ \hline 1.06 \times 10^4 \text{ Pa} \end{array}}$$

Part b)

Systolic:

$$120 \text{ mmHg} \times \frac{1 \text{ atm}}{760 \text{ mmHg}} = 0.158 \text{ atm}$$

Diastolic:

$$80 \text{ mmHg} \times \frac{1 \text{ atm}}{760 \text{ mmHg}} = 0.105 \text{ atm}$$

Blood pressure:

$$\boxed{\begin{array}{c} 0.158 \text{ atm} \\ \hline 0.105 \text{ atm} \end{array}}$$

Part c)

Systolic:

$$120 \text{ mmHg} \times \frac{14.7 \text{ psi}}{760 \text{ mmHg}} = 2.32 \text{ psi}$$

Diastolic:

$$80 \text{ mmHg} \times \frac{14.7 \text{ psi}}{760 \text{ mmHg}} = 1.55 \text{ psi}$$

Blood pressure:

$$\boxed{\begin{array}{c} 2.32 \text{ psi} \\ \hline 1.55 \text{ psi} \end{array}}$$

Part d) Blood pressure is reported as a gauge pressure. When you're cut, blood comes out of your arteries; the air doesn't rush in.

REFLECT

If the blood pressures reported were absolute pressures, these would be much smaller than atmospheric pressure. In that case, the pressure difference between the vessels and the outside air would compress all of our blood vessels shut, which luckily does not happen.

Get Help: P'Cast 11.5 – Diver's Rule of Thumb

11.93

SET UP

A syringe has an inner diameter of 0.6×10^{-3} m. A nurse uses this syringe to inject fluid into a patient's artery where the blood pressure is 140/100. The minimum force the nurse needs to apply to the syringe during the injection is equal to the diastolic blood pressure multiplied by the cross-sectional area of the syringe.

SOLVE

$$p = \frac{F_\perp}{A}$$

$$F_\perp = pA = \left(100 \text{ mmHg} \times \frac{1.01 \times 10^5 \text{ Pa}}{760 \text{ mmHg}}\right)(\pi)\left(\frac{0.6 \times 10^{-3} \text{ m}}{2}\right)^2 = \boxed{0.004 \text{ N}}$$

REFLECT

The minimum force required corresponds to the minimum pressure in the artery, which is the diastolic pressure.

Get Help: P'Cast 11.5 – Diver's Rule of Thumb

Chapter 12
Oscillations

Conceptual Questions

12.1 Simple harmonic motion is oscillatory motion in which the displacement of the object is proportional, but opposite in direction, to the force on the object. Oscillatory motion not only includes simple harmonic motion, but it also includes circular motion, decaying oscillations, and oscillations that do not have a sinusoidal time dependence. An example of simple harmonic motion is the movement of a mass on an ideal spring. An example of oscillatory motion that is not simple harmonic motion is the up-and-down motion of a yo-yo.

12.5 Breathing rate (breaths per minute) is a frequency. The period is its reciprocal.

12.11 The force of gravity depends on elevation and determines the period of the pendulum. Measuring the period and length of the pendulum will yield enough information to measure the acceleration due to gravity; doing so very precisely would enable useful comparison to established values to estimate the elevation.

Get Help: Picture It – Period of a Simple Pendulum

12.13 The frequency of the driving force is just how often the applied force repeats. That depends on things outside the oscillator. The natural frequency of the oscillator is the frequency at which it oscillates most readily or the frequency at which it will oscillate if displaced from equilibrium and released.

Multiple-Choice Questions

12.15 **B** ($x = 0$). The kinetic energy of the block is a maximum when the spring is at its equilibrium position.

Get Help: Interactive Exercise – Block and Spring

12.19 **A** (The period will increase).

$$T_1 = \frac{2\pi}{\omega_1} = 2\pi\sqrt{\frac{m_1}{k}}$$

$$T_2 = \frac{2\pi}{\omega_2} = 2\pi\sqrt{\frac{m_2}{k}} = 2\pi\sqrt{\frac{(2m_1)}{k}} = \sqrt{2}\left(2\pi\sqrt{\frac{m_1}{k}}\right) = \sqrt{2}T_1$$

Get Help: Interactive Exercise – Block and Spring

Estimation/Numerical Analysis

12.27 Swings are typically around 2 m in length, so we get a period of around 3 s.

$$T = 2\pi\sqrt{\frac{L}{g}} = 2\pi\sqrt{\frac{2\text{ m}}{9.80\frac{\text{m}}{\text{s}^2}}} \approx 3\text{ s}$$

Get Help: P'Cast 12.1 – Frequency and Period

12.33 A 0.01-g fly clinging to the bottom of a pendulum that includes a rod with a mass of 1 kg. The bob is a disk with a mass of 1 kg and a radius of 10 cm. It is centered at 1 m from the pivot of the pendulum. We can use the following equations to calculate the frequency of the pendulum and the pendulum + fly:

$$f_{\text{pendulum}} = \frac{1}{2\pi}\sqrt{\frac{mgh}{I}} = \frac{1}{2\pi}\sqrt{\frac{m_{\text{rod}}g\left(\frac{L}{2}\right) + m_{\text{disc}}gL}{\frac{1}{3}m_{\text{rod}}L^2 + \frac{1}{2}m_{\text{disc}}r^2 + m_{\text{disc}}L^2}}$$

$$f_{\text{pendulum}} = \frac{1}{2\pi}\sqrt{\frac{(1\text{ kg})\left(9.80\frac{\text{m}}{\text{s}^2}\right)\left(\frac{1\text{ m}}{2}\right) + (1\text{ kg})\left(9.80\frac{\text{m}}{\text{s}^2}\right)(1\text{ m})}{\frac{1}{3}(1\text{ kg})(1\text{ m})^2 + \frac{1}{2}(1\text{ kg})(0.1\text{ m})^2 + (1\text{ kg})(1\text{ m})^2}}$$

$$= 0.527468648837\text{ Hz}$$

$$f_{\text{pendulum + fly}} = \frac{1}{2\pi}\sqrt{\frac{mgh}{I}} = \frac{1}{2\pi}\sqrt{\frac{m_{\text{rod}}g\left(\frac{L}{2}\right) + m_{\text{disc}}gL + m_{\text{fly}}g(L+r)}{\frac{1}{3}m_{\text{rod}}L^2 + \frac{1}{2}m_{\text{disc}}r^2 + m_{\text{disc}}L^2 + m_{\text{fly}}(L+r)^2}}$$

$$f_{\text{pendulum + fly}} = \frac{1}{2\pi}\sqrt{\frac{(1\text{ kg})\left(9.80\frac{\text{m}}{\text{s}^2}\right)\left(\frac{1\text{ m}}{2}\right) + (1\text{ kg})\left(9.80\frac{\text{m}}{\text{s}^2}\right)(1\text{ m}) + (0.00001\text{ kg})\left(9.80\frac{\text{m}}{\text{s}^2}\right)(1.1\text{ m})}{\frac{1}{3}(1\text{ kg})(1\text{ m})^2 + \frac{1}{2}(1\text{ kg})(0.1\text{ m})^2 + (1\text{ kg})(1\text{ m})^2 + (0.00001\text{ kg})(1.1\text{ m})^2}}$$

$$= 0.5274681985\text{ Hz}$$

$$\Delta f = f_{\text{pendulum}} - f_{\text{pendulum + fly}} = 0.527468648837\text{ Hz} - 0.5274681985\text{ Hz} \approx 4 \times 10^{-7}\text{ Hz}$$

The fly landing on the pendulum would reduce the frequency of the pendulum by about 4×10^{-7} Hz.

Get Help: Picture It – Period of a Simple Pendulum
P'Cast 12.5 – Changing a Pendulum

12.37 On average, the period looks to be about 4 years.

Get Help: P'Cast 12.1 – Frequency and Period

Problems

12.41

SET UP

An object undergoing simple harmonic motion has a frequency of $f = 15$ Hz. The period of the object's motion is equal to the reciprocal of its frequency. Multiplying the frequency of the motion by 2 min will give us the number of cycles the object undergoes in that time period.

SOLVE

Part a)

$$T = \frac{1}{f} = \frac{1}{15 \text{ Hz}} = \boxed{0.067 \text{ s}}$$

Part b)

$$N_{\text{cycles}} = 2 \text{ min} \times \frac{60 \text{ s}}{1 \text{ min}} \times \frac{15 \text{ cycles}}{1 \text{ s}} = \boxed{1.8 \times 10^3 \text{ cycles}}$$

REFLECT

A frequency of 15 Hz is reasonably fast for a mass on the end of a spring.

Get Help: P'Cast 12.1 – Frequency and Period

12.47

SET UP

A 0.25-kg object attached to a spring oscillates on a frictionless horizontal table with a frequency of $f = 4.00$ Hz and an amplitude of $A = 0.200$ m. Although we were not given it explicitly, we can rewrite the spring constant in terms of the frequency and mass. The maximum potential energy of the system occurs when the kinetic energy is equal to zero, which means the maximum potential energy is equal to the total energy of the harmonic oscillator. We can then divide this quantity by two and solve for x to solve for the displacement of the object when the potential energy is one-half of the maximum. Finally, we can plug $x = 0.100$ m into the elastic potential energy to find the potential energy of the object at that displacement.

SOLVE

Part a)

$$E = K + U_{\text{spring}} = \frac{1}{2}kA^2$$

$$U_{\text{spring,max}} = \frac{1}{2}kA^2 - K = \frac{1}{2}kA^2 - 0 = \frac{1}{2}kA^2 = \frac{1}{2}\omega^2 mA^2 = \frac{1}{2}(2\pi f)^2 mA^2 = 2\pi^2 m f^2 A^2$$

$$= 2\pi^2(0.25 \text{ kg})(4.00 \text{ Hz})^2(0.200 \text{ m})^2 = \boxed{3.2 \text{ J}}$$

Part b)

$$U_{\text{spring,half max}} = \frac{1}{2}kx^2 = \frac{1}{2}\left(\frac{1}{2}kA^2\right)$$

$$x = \frac{A}{\sqrt{2}} = \frac{20.0 \text{ cm}}{\sqrt{2}} = \boxed{14.1 \text{ cm}}$$

Part c)

$$U_{spring}(x) = \frac{1}{2}kx^2 = \frac{1}{2}\omega^2 mx^2 = \frac{1}{2}(2\pi f)^2 mx^2 = 2\pi^2 mf^2 x^2$$

$$U_{spring}(x = 0.100 \text{ m}) = 2\pi^2(0.25 \text{ kg})(4.00 \text{ Hz})^2(0.100 \text{ m})^2 = \boxed{0.79 \text{ J}}$$

REFLECT
The displacement where the potential energy is one-half of the maximum value does *not* occur at one-half the amplitude.

12.49

SET UP
The equation of motion for a given simple harmonic oscillator is $x = A\cos(\omega t)$. The velocity of this oscillator is given by Equation 12-14, $v_x = -\omega A \sin(\omega t)$. The potential energy of a simple harmonic oscillator is given by $U(t) = \frac{1}{2}kx^2$ and the kinetic energy is given by $K(t) = \frac{1}{2}mv_x^2$. The plots of $U(t)$ and $K(t)$ should look like those for $\cos^2(\omega t)$ and $\sin^2(\omega t)$, respectively.

SOLVE
Part a)

$$U(t) = \frac{1}{2}kx^2 = \frac{1}{2}k[A\cos(\omega t)]^2 = \frac{1}{2}kA^2\cos^2(\omega t)$$

For simplicity, the vertical axis of the following plot is $\dfrac{U}{\left(\frac{1}{2}kA^2\right)}$, and the horizontal axis is ωt:

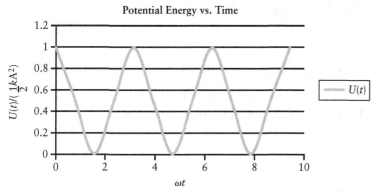

Figure 12-1 Problem 49

Part b)

$$v_x = -\omega A \sin(\omega t)$$

Part c)

$$K(t) = \frac{1}{2}mv^2 = \frac{1}{2}m[-\omega A \sin(\omega t)]^2 = \frac{1}{2}m\omega^2 A^2 \sin^2(\omega t) = \frac{1}{2}kA^2 \sin^2(\omega t)$$

For simplicity, the vertical axis of the following plot is $\dfrac{U}{\left(\frac{1}{2}kA^2\right)}, \dfrac{K}{\left(\frac{1}{2}kA^2\right)}$ and the horizontal axis is ωt:

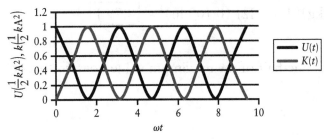

Figure 12-2 Problem 49

REFLECT
Although U and K oscillate in time, their sum is constant for all time.

12.53

SET UP
A simple pendulum completes 14 cycles in 25.0 s. We can use the definition of the period of a simple pendulum $T = 2\pi\sqrt{\dfrac{L}{g}}$ and solve for its length.

SOLVE

$$T = 2\pi\sqrt{\dfrac{L}{g}}$$

$$L = g\left(\dfrac{T}{2\pi}\right)^2 = \left(9.80\dfrac{m}{s^2}\right)\left(\dfrac{\left(\dfrac{25.0\ s}{14\ cycles}\right)}{2\pi}\right)^2 = \boxed{0.792\ m}$$

REFLECT
The period of a pendulum is the time it takes to make one full oscillation.

Get Help: Picture It – Period of a Simple Pendulum
P'Cast 12.5 – Changing a Pendulum

12.57

SET UP
A simple pendulum has a length of $L = 0.50$ m and oscillates with an amplitude of 8°. From these data we can calculate the period of the motion and write down the equation of motion for the pendulum. The time it takes the pendulum to swing from +8° to −8° is half of a period. To find the time interval between +4° and −4°, we will need to find the times at which the pendulum is at each of these angular positions; we can do this by solving the equation of motion.

SOLVE
Period:

$$T = 2\pi\sqrt{\frac{L}{g}} = 2\pi\sqrt{\frac{0.50 \text{ m}}{\left(9.80\frac{\text{m}}{\text{s}^2}\right)}} = 1.42 \text{ s}$$

Equation of motion:

$$\theta(t) = \theta_0\cos(\omega t) = \theta_0\cos\left(\frac{2\pi t}{T}\right) = (8°)\cos\left(\left(4.42\frac{\text{rad}}{\text{s}}\right)t\right)$$

Time interval between $+8°$ and $-8°$:

$$\frac{T}{2} = \boxed{0.71 \text{ s}}$$

Time interval between $+4°$ and $-4°$:

$$4° = (8°)\cos\left(\left(4.42\frac{\text{rad}}{\text{s}}\right)t_1\right)$$

$$t_1 = \frac{\cos^{-1}\left(\frac{4°}{8°}\right)}{\left(4.42\frac{\text{rad}}{\text{s}}\right)} = 0.237 \text{ s}$$

$$-4° = (8°)\cos\left(\left(4.42\frac{\text{rad}}{\text{s}}\right)t_2\right)$$

$$t_2 = \frac{\cos^{-1}\left(\frac{-4°}{8°}\right)}{\left(4.42\frac{\text{rad}}{\text{s}}\right)} = 0.474 \text{ s}$$

$$\Delta t = t_2 - t_1 = (0.474 \text{ s}) - (0.237 \text{ s}) = \boxed{0.2 \text{ s}}$$

REFLECT
The time interval between $+4°$ and $-4°$ is one-third of the period.

Get Help: Picture It – Period of a Simple Pendulum
P'Cast 12.5 – Changing a Pendulum

12.61

SET UP
A solid sphere of mass m and radius $r = 0.050$ m is suspended from an eyelet attached to its surface. The sphere is displaced slightly from equilibrium and set into simple harmonic motion. The period of a physical pendulum is related to the moment of inertia of the pendulum. Since the center of mass of the sphere is located a distance $h = r$ from the pivot point, we will need to use the parallel-axis theorem. As a reminder, the moment of inertia of a sphere is $\frac{2}{5}mr^2$.

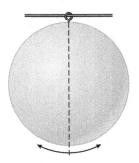

Figure 12-3 Problem 61

SOLVE

$$T = 2\pi\sqrt{\frac{I}{mgh}} = 2\pi\sqrt{\frac{I_{\text{sphere}} + I_{\text{from axis}}}{mgh}} = 2\pi\sqrt{\frac{\left(\frac{2}{5}mr^2\right) + mr^2}{mgr}} = 2\pi\sqrt{\frac{7r}{5g}}$$

$$= 2\pi\sqrt{\frac{7(0.050\text{ m})}{5\left(9.80\frac{\text{m}}{\text{s}^2}\right)}} = \boxed{0.53\text{ s}}$$

REFLECT

The density of the material is irrelevant since the mass cancels out in our expression for the period of the oscillation.

Get Help: P'Cast 12.6 – An Oscillating Rod
P'Cast 12.7 – Moment of Inertia of a Human Leg

12.65

SET UP

An oscillating system has a natural frequency of $\omega_0 = 50$ rad/s and a damping coefficient of $b = 2.0$ kg/s. The system is driven by an oscillating force, $F(t) = (100\text{ N})\cos\left(\left(50\frac{\text{rad}}{\text{s}}\right)t\right)$.
From this expression, we can see that the frequency of the applied force is equal to the natural frequency of the system. The amplitude of a damped, driven oscillator is
$A = \dfrac{F_0}{\sqrt{m^2(\omega_0^2 - \omega^2)^2 + b^2\omega^2}}$.

SOLVE

$$A = \frac{F_0}{\sqrt{m^2(\omega_0^2 - \omega^2)^2 + b^2\omega^2}} = \frac{F_0}{\sqrt{m^2(\omega_0^2 - \omega^2)^2 + b^2\omega_0^2}} = \frac{F_0}{b\omega_0}$$

$$= \frac{100\text{ N}}{\left(2.0\frac{\text{kg}}{\text{s}}\right)\left(50\frac{\text{rad}}{\text{s}}\right)} = \boxed{1\text{ m}}$$

REFLECT

This is the maximum amplitude of the system's motion because the applied force oscillates at the natural frequency of the system.

General Problems

12.67

SET UP

The acceleration of an object ($m = 0.025$ kg) that exhibits simple harmonic motion is given by $a(t) = \left(10\dfrac{\text{m}}{\text{s}^2}\right)\cos\left(\pi t + \dfrac{\pi}{2}\right)$. We can compare Equations 12-14 and 12-15 from the text to determine the equation describing the velocity of the object as a function of time. Although the velocity and acceleration expressions differ in the trigonometric function used (sine vs. cosine, respectively) and their amplitudes (by a factor of ω), the argument of the trigonometric functions are the same. We can then calculate the velocity at $t = 2$ s since the velocity is a maximum at $t = 0$.

SOLVE

General form of the acceleration as a function of time for simple harmonic motion:

$$a_x = -\omega^2 A \cos(\omega t + \phi)$$

General form of the velocity as a function of time for simple harmonic motion:

$$v_x = -\omega A \sin(\omega t + \phi)$$

By comparison, the amplitude of the velocity expression is equal to the amplitude of the acceleration divided by the angular frequency. Therefore,

$$v_x = \left(\dfrac{10}{\pi}\dfrac{\text{m}}{\text{s}}\right)\sin\left(\pi t + \dfrac{\pi}{2}\right)$$

Velocity at $t = 2$ s:

$$v_x(t = 2\text{ s}) = \left(\dfrac{10}{\pi}\dfrac{\text{m}}{\text{s}}\right)\sin\left(\pi(2) + \dfrac{\pi}{2}\right) = \dfrac{10}{\pi}\dfrac{\text{m}}{\text{s}} = \boxed{3\dfrac{\text{m}}{\text{s}}}$$

REFLECT

The general equation of the acceleration in terms of the period of the oscillation is $a(t) = a_0 \cos\left(\dfrac{2\pi t}{T} + \phi\right)$. The period of the oscillation in this problem is $T = 2$ s, which means the velocity at $t = 0$ s will be equal to the velocity at $t = 2$ s.

12.69

SET UP

The equation describing the motion of a particular simple harmonic oscillator is $x = (0.15\text{ m})\cos\left(\pi t + \dfrac{\pi}{3}\right)$. We can compare Equations 12-12, 12-14, and 12-15 from the text to determine the equations describing the velocity and acceleration of the oscillator as a function of time. Although the position, velocity, and acceleration expressions differ in the trigonometric function used (cosine, sine, and cosine, respectively) and their amplitudes (by a factor of ω), the arguments of the trigonometric functions are the same. We can then calculate the velocity at $t = 1$ s and the acceleration at $t = 2$ s.

128 Chapter 12 Oscillations

SOLVE
General form of the position as a function of time for simple harmonic motion:
$$x = A\cos(\omega t + \phi)$$
General form of the velocity as a function of time for simple harmonic motion:
$$v_x = -\omega A \sin(\omega t + \phi)$$
General form of the acceleration as a function of time for simple harmonic motion:
$$a_x = -\omega^2 A \cos(\omega t + \phi)$$
From the given equation, we can conclude:
$$A = 0.15 \text{ m} \qquad \omega = \pi \qquad \phi = \frac{\pi}{3}$$
By comparison, the amplitude of the velocity expression is equal to the amplitude of the position multiplied by $-\omega$, and the amplitude of the acceleration expression is equal to the amplitude of the position multiplied by $-\omega^2$. Therefore,
$$v_x = -\left(0.15\pi \frac{\text{m}}{\text{s}}\right) \sin\left(\pi t + \frac{\pi}{3}\right)$$
$$a_x = -\left(0.15\pi^2 \frac{\text{m}}{\text{s}^2}\right) \cos\left(\pi t + \frac{\pi}{3}\right)$$

Part a)
Velocity at $t = 1$ s (SI units):
$$v_x(t=1) = -(0.15\pi)\sin\left(\pi(1) + \frac{\pi}{3}\right) = \boxed{0.41 \frac{\text{m}}{\text{s}}}$$

Part b)
Acceleration at $t = 2$ s (SI units):
$$a_x(t=2) = -(0.15\pi^2)\cos\left(\pi(2) + \frac{\pi}{3}\right) = \boxed{-0.74 \frac{\text{m}}{\text{s}^2}}$$

REFLECT
The position and acceleration of a simple harmonic oscillator are exactly 180° out of phase with one another. This is why they have the same trigonometric function, but the acceleration expression has a leading negative sign.

12.73

SET UP
A damped oscillator shows a reduction of 30% in amplitude after two periods. We can use the amplitude of a damped oscillator as a function of time, $A(t) = A_0 e^{-(b/2m)t}$, to find the reduction in amplitude and the mechanical energy loss after one period.

SOLVE

Amplitude after one period:

$$A(t) = Ae^{-(b/2m)t}$$

$$\frac{A_{2T}}{A_0} = e^{-(b/2m)(2T)} = \left(e^{-(b/2m)(T)}\right)^2 = \left(\frac{A_T}{A_0}\right)^2 = 0.7$$

Percent loss in mechanical energy per cycle:

$$\left(\frac{E_T - E_0}{E_0}\right) = \frac{\left(\frac{1}{2}kA_T^2\right) - \left(\frac{1}{2}kA_0^2\right)}{\left(\frac{1}{2}kA_0^2\right)} = \left(\frac{A_T}{A_0}\right)^2 - 1 = (0.7) - 1 = -0.3$$

There is a $\boxed{30\% \text{ loss}}$ of mechanical energy per cycle.

REFLECT

The change in mechanical energy of the system is equal to the work done by all nonconservative forces on the system.

Get Help: Picture It – Damped Oscillations

12.83

SET UP

Two springs with spring constants k_1 and k_2, respectively, are attached end-to-end to a box of mass M. We can find the effective spring constant of this setup by comparing the two-spring setup with an equivalent one-spring setup. If the box is moved a distance Δx, each spring will stretch a different amount (Δx_1 and Δx_2) but their sum must be Δx. In the figure, we see that only spring 2 is attached to the mass, which means only the force due to spring 2 acts on the mass. Newton's third law tells us that the magnitude of the force of spring 1 on spring 2 must equal the magnitude of the force of spring 2 on spring 1: $k_1\Delta x_1 = k_2\Delta x_2$; we can use this fact to eliminate Δx_1 from our equation. The period of the motion of the box is equal to $T = 2\pi\sqrt{\frac{M}{k_{\text{eff}}}}$.

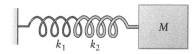

Figure 12-4 Problem 83

SOLVE

$$F_{1x} = k_1\Delta x_1$$

$$F_{2x} = k_2\Delta x_2$$

$$k_1\Delta x_1 = k_2\Delta x_2$$

$$\Delta x_1 = \frac{k_2 \Delta x_2}{k_1}$$

$$\Delta x = \Delta x_1 + \Delta x_2 = \left(\frac{k_2 \Delta x_2}{k_1}\right) + \Delta x_2 = \left(\frac{k_2}{k_1} + 1\right)\Delta x_2$$

$$\Delta x_2 = \frac{\Delta x}{\left(\frac{k_2}{k_1} + 1\right)}$$

Effective spring constant:

$$F_{2x} = F_{\text{eff},x}$$

$$k_2 \Delta x_2 = k_{\text{eff}} \Delta x$$

$$k_2 \left(\frac{\Delta x}{\left(\frac{k_2}{k_1} + 1\right)}\right) = k_{\text{eff}} \Delta x$$

$$k_{\text{eff}} = \frac{k_2}{\left(\frac{k_2}{k_1} + 1\right)} = \frac{k_2}{\left(\frac{k_2 + k_1}{k_1}\right)} = \frac{k_1 k_2}{(k_2 + k_1)}$$

Period:

$$T = \frac{2\pi}{\omega} = 2\pi\sqrt{\frac{M}{k_{\text{eff}}}} = \boxed{2\pi\sqrt{\frac{M(k_1 + k_2)}{k_1 k_2}} = 2\pi\sqrt{M\left(\frac{1}{k_1} + \frac{1}{k_2}\right)}}$$

REFLECT

The springs in this configuration are described as being in "series." The effective spring constant in this case is *smaller* than the individual spring constants.

Get Help: Interactive Exercise – Block and Spring

Chapter 13
Waves

Conceptual Questions

13.3 A very few of them might be the same molecules, especially if he is close by. The air molecules collide many times on the way from place to place. The collisions transmit the wave motion consisting of a wave pattern of alternating compressions and rarefactions of the air, without requiring the individual molecules to go the full distance.

13.7 Part a) The wave speed depends entirely on the medium so wave speed increases as the wave passes from air into water.

Part b) The frequency is determined by the actual vibrations of the source, so the frequency does not change.

Part c) The wavelength depends on changes in the wave speed, so the wavelength increases, too.

Get Help: Interactive Exercise – Tension
Interactive Exercise – Slinky

13.11 The frequency counts the number of oscillations per time. The angular frequency counts the oscillations per time in units of 2π; one complete oscillation is equivalent to 2π rad. The angular frequency is more natural from a mathematical point of view.

13.19 Part a) Intensity is the power per unit area that is emitted by a source of sound. It has nothing to do with the reception of the sound through the process of "hearing." The SI units are W/m^2. Sound level is a mathematical relationship that puts the intensity onto a logarithmic scale. The units are decibels (dB). Loudness is the physiological response to sound waves. Different frequencies with the same intensity will be more sensitively heard by humans. Power is the energy per unit time emitted by the source.

Part b) Intensity depends on the inverse of the distance squared, so the intensity increases as the source of sound moves closer.

Part c) Sound level depends on the log of intensity, so sound level increases as the source of sound moves closer.

Part d) The total power emitted by the source does not change as it moves closer to the observer.

Multiple-Choice Questions

13.23 D (5).

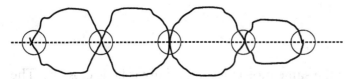

Figure 13-1 Problem 23

13.27 D (The sound. intensity drops to 1/9 its original value). The sound intensity is inversely proportional to the square of the distance.

Estimation/Numerical Analysis

13.31 Sound with a frequency of 3000 Hz travels through air at a speed of 343 m/s. The wavelength of the sound is:

$$v_p = f\lambda$$

$$\lambda = \frac{v_p}{f} = \frac{343\,\frac{m}{s}}{3000\text{ Hz}} = 0.1143\text{ m} = 11.43\text{ cm}$$

If the voice box is considered a closed organ pipe, the length of the voice box is given by:

$$L = \frac{1}{4}\lambda = \left(\frac{1}{4}\right)(11.43\text{ cm}) = 2.9\text{ cm}$$

Get Help: P'Cast 13.7 – An Amorous Frog

13.35 We can measure the pitch of the fundamental compared to a known standard (like the note A 440 Hz) and measure the length of the string. The speed is twice the length divided by the frequency.

Problems

13.41

SET UP

A wave on a string propagates at $v_p = 22$ m/s. The frequency of the wave is $f = 24$ Hz. We can use the speed of the wave and the frequency to calculate the wavelength. The wavenumber is $k = \frac{2\pi}{\lambda}$.

SOLVE
Wavelength:

$$v_p = f\lambda$$

$$\lambda = \frac{v_p}{f} = \frac{\left(22\frac{m}{s}\right)}{24 \text{ Hz}} = \boxed{0.92 \text{ m} = 92 \text{ cm}}$$

Wavenumber:

$$k = \frac{2\pi}{\lambda} = \frac{2\pi}{\left(\frac{v_p}{f}\right)} = \frac{2\pi f}{v_p} = \frac{2\pi(24 \text{ Hz})}{\left(22\frac{m}{s}\right)} = \boxed{6.9\frac{\text{rad}}{\text{m}}}$$

REFLECT
Be careful when rounding intermediate values. You should carry along one more significant figure than you are allowed in your final answer throughout your calculations.

Get Help: P'Cast 13.1 – Wave Speed on a Sperm's Flagellum

13.45

SET UP

A wave on a string is described by $y(x, t) = (0.5 \text{ m})\cos\left[\left(1.0\frac{\text{rad}}{\text{m}}\right)x - \left(10\frac{\text{rad}}{\text{s}}\right)t\right]$.

We can compare this equation to the general form of a transverse wave, $y(x,t) = A\cos(kx - \omega t + \phi)$, along with the definitions of the wavenumber and angular frequency in order to determine the frequency, the wavelength, and the speed of the wave.

SOLVE
Part a)

$$\omega = 2\pi f$$

$$f = \frac{\omega}{2\pi} = \frac{\left(10\frac{\text{rad}}{\text{s}}\right)}{2\pi} = \boxed{2 \text{ Hz}}$$

Part b)

$$\lambda = \frac{2\pi}{k} = \frac{2\pi}{1.0 \text{ m}^{-1}} = \boxed{6.3 \text{ m}}$$

Part c)

$$v_p = \frac{\omega}{k} = \frac{\left(10\frac{\text{rad}}{\text{s}}\right)}{1.0 \text{ m}^{-1}} = \boxed{1 \times 10^1 \frac{\text{m}}{\text{s}}}$$

REFLECT
We could have also used $v_p = \lambda f$ to find the speed, but it's better to use the values provided rather than intermediate values that we've calculated.

Get Help: P'Cast 13.1 – Wave Speed on a Sperm's Flagellum

13.49

SET UP

We are given plots of a wave at $t = 0$ and $x = 0$. Since the plots look like sine functions, we will use this in our mathematical description. We can read the amplitude, wavelength, and period directly from the graphs. The wavenumber and angular frequency can be calculated from the wavelength and period. Because the wave starts at $y = 0$, the phase is also equal to 0.

$t = 0$ s:

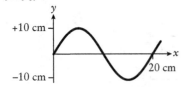

$x = 0$ m:

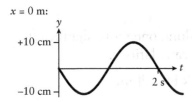

Figure 13-2 Problem 49

SOLVE

Amplitude, $A = 0.10$ m

Wavelength, $\lambda = 0.20$ m

Period, $T = 2$ s

Wavenumber:

$$k = \frac{2\pi}{\lambda} = \frac{2\pi}{(0.20 \text{ m})} = 10\pi \text{ m}^{-1}$$

Angular frequency:

$$\omega = \frac{2\pi}{T} = \frac{2\pi}{2 \text{ s}} = \pi \frac{\text{rad}}{\text{s}}$$

Mathematical description:

$$\boxed{y(x, t) = 0.10 \sin(10\pi x - \pi t)} \text{ (SI units)}$$

REFLECT

The plot y versus t looks like $-\sin(\omega t)$ because sine is an odd function; that is, $\sin(-\omega t) = -\sin(\omega t)$.

13.51

SET UP

A string has a mass of 5.0×10^{-3} kg and length of 2.2 m. The string is pulled taut with a tension of 78 N. The propagation speed of a transverse wave on a string is given by $v_p = \sqrt{\frac{F}{\mu}}$, where F is the tension and μ is the linear mass density.

SOLVE

$$v_p = \sqrt{\frac{F}{\mu}} = \sqrt{\frac{78 \text{ N}}{\left(\frac{5.0 \times 10^{-3} \text{ kg}}{2.2 \text{ m}}\right)}} = \boxed{1.9 \times 10^2 \frac{\text{m}}{\text{s}}}$$

REFLECT

The propagation speed is how fast the wave travels down the rope, not how fast a small piece of the rope moves up and down.

Get Help: Interactive Exercise – Tension
Interactive Exercise – Slinky
P'Cast 13.3 – Tension in a Sperm's Flagellum

13.55

SET UP

The bulk modulus and density of water are $B = 2.2 \times 10^9$ Pa and $\rho = 1000$ kg/m³, respectively. We can use this information to find the speed of sound in water using $v_p = \sqrt{\frac{B}{\rho}}$.

SOLVE

$$v_p = \sqrt{\frac{B}{\rho}} = \sqrt{\frac{2.2 \times 10^9 \text{ Pa}}{\left(1000 \frac{\text{kg}}{\text{m}^3}\right)}} = \boxed{1.5 \times 10^3 \frac{\text{m}}{\text{s}}}$$

REFLECT

For comparison, the speed of sound in air is 343 m/s.

Get Help: Interactive Exercise – Tension
Interactive Exercise – Slinky
P'Cast 13.4 – Propagation Speeds

13.59

SET UP

We are shown four different scenarios where square waveforms of different widths and amplitudes are approaching one another (see figure). When they overlap in time and space, they will interfere with one another. If the waveforms are both upright, they will constructively interfere; if the waveforms are inverted relative to one another, they will destructively interfere.

136 Chapter 13 Waves

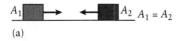
(a)

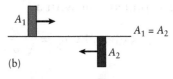
(b)

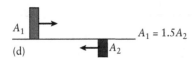
(c)

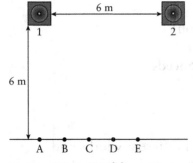
(d)

Figure 13-3 Problem 59

SOLVE

Part a) The resulting waveform will have the same width as the incoming waveforms but an amplitude of $A = A_1 + A_2 = \boxed{2A_1}$.

Part b) The resulting waveform will have an amplitude of $A = A_1 - A_2 = \boxed{0}$.

Part c) The resulting waveform will have the same width as the incoming waveforms but will have an amplitude of $A = A_1 + A_2 = \boxed{2.5A_1}$.

Part d) The resulting waveform will have the same width as the incoming waveforms but will have an amplitude of $A = A_1 - A_2 = \boxed{A_1/2}$.

REFLECT

The waveforms will pass through one another unaffected. Only in the exact moment when they are coincident will we observe another square waveform with the amplitudes we found.

13.61

Figure 13-4 Problem 61

SET UP

Two speakers are 6 m apart and playing the same tone ($f = 171.5$ Hz) in phase. The speed of sound ($v_p = 343$ m/s) divided by the frequency of the tone will give us its wavelength. In order to determine which of the labeled points (A–E) will experience fully constructive interference, we need to use geometry to determine the path length difference between the two waves. The path length for each wave is equal to the hypotenuse of a triangle made by the 6-m distance between the speakers and the line of points and the distance to the point. Keep in mind that each of the labeled points is 1 m from its neighbors. If the path length difference divided by the wavelength of the tone is an integer, then that point will experience fully constructive interference; if not, there will be partial destructive interference.

SOLVE

Wavelength:

$$v_p = f\lambda$$

$$\lambda = \frac{v_p}{f} = \frac{\left(343\frac{\text{m}}{\text{s}}\right)}{171.5 \text{ Hz}} = 2.00 \text{ m}$$

Point A)

$$\Delta_{pl} = D_2 - D_1 = (\sqrt{(6 \text{ m})^2 + (6 \text{ m})^2}) - (6 \text{ m}) = 2.49 \text{ m}$$

$$\Delta_{pl} = n\lambda$$

$$n = \frac{\Delta_{pl}}{\lambda} = \frac{2.49 \text{ m}}{2.00 \text{ m}} = 1.24$$

Since n is not an integer, point A will not experience constructive interference.

Point B)

$$\Delta_{pl} = D_2 - D_1 = (\sqrt{(5 \text{ m})^2 + (6 \text{ m})^2}) - (\sqrt{(1 \text{ m})^2 + (6 \text{ m})^2}) = 1.73 \text{ m}$$

$$\Delta_{pl} = n\lambda$$

$$n = \frac{\Delta_{pl}}{\lambda} = \frac{1.73 \text{ m}}{2.00 \text{ m}} = 0.865$$

Since n is not an integer, point B will not experience constructive interference.

Point C)

$$\Delta_{pl} = D_2 - D_1 = (\sqrt{(4 \text{ m})^2 + (6 \text{ m})^2}) - (\sqrt{(2 \text{ m})^2 + (6 \text{ m})^2}) = 0.887 \text{ m}$$

$$\Delta_{pl} = n\lambda$$

$$n = \frac{\Delta_{pl}}{\lambda} = \frac{0.887 \text{ m}}{2.00 \text{ m}} = 0.444$$

Since n is not an integer, point C will not experience constructive interference.

Point D)
$$\Delta_{pl} = D_2 - D_1 = (\sqrt{(3\text{ m})^2 + (6\text{ m})^2}) - (\sqrt{(3\text{ m})^2 + (6\text{ m})^2}) = 0\text{ m}$$

$$\Delta_{pl} = n\lambda$$

$$n = \frac{\Delta_{pl}}{\lambda} = \frac{0\text{ m}}{2.00\text{ m}} = 0$$

Since n is an integer, point D will experience constructive interference.

Point E)
$$\Delta_{pl} = D_2 - D_1 = (\sqrt{(2\text{ m})^2 + (6\text{ m})^2}) - (\sqrt{(4\text{ m})^2 + (6\text{ m})^2}) = -0.887\text{ m}$$

$$\Delta_{pl} = n\lambda$$

$$n = \frac{\Delta_{pl}}{\lambda} = \frac{-0.887\text{ m}}{2.00\text{ m}} = -0.444$$

Since n is not an integer, point E will not experience constructive interference.

REFLECT

The speakers are symmetric about a vertical line through point D, which means the distances from each speaker to that point will be equal.

13.65

SET UP

A string with a length $L = 1.25$ m and mass $m = 0.0548$ kg has a tension of $F = 200.0$ N. It is fixed at both ends. We want to calculate the frequencies for the first four harmonics. The speed of the waves on the string is equal to $v_p = \sqrt{\frac{F}{\mu}}$, where μ is the linear mass density.

SOLVE

Part a)

To calculate the allowed frequencies, first solve for frequency:

$$v_p = f\lambda$$

$$f = \frac{v_p}{\lambda}$$

Note that the wavelength of a standing wave on a string is given by:

$$L = \frac{n\lambda}{2}$$

$$\lambda = \frac{2L}{n}$$

Combining these equations gives:

$$f = \frac{v_p}{\lambda} = \frac{v_p n}{2L}$$

We can substitute the expression relating tension and linear mass density of the string to obtain the standing wave frequencies:

$$f = \frac{v_p n}{2L} = \frac{n}{2L}\sqrt{\frac{F}{\mu}} = \frac{n}{2L}\sqrt{\frac{F}{\left(\frac{m}{L}\right)}} = \frac{n}{2}\sqrt{\frac{F}{Lm}}$$

First harmonic ($n = 1$):

$$f_1 = \frac{1}{2}\sqrt{\frac{200.0 \text{ N}}{(1.25 \text{ m})(0.0548 \text{ kg})}} = \boxed{27.0 \text{ Hz}}$$

Second harmonic ($n = 2$):

$$f_1 = \sqrt{\frac{200.0 \text{ N}}{(1.25 \text{ m})(0.0548 \text{ kg})}} = \boxed{54.0 \text{ Hz}}$$

Third harmonic ($n = 3$):

$$f_1 = \frac{3}{2}\sqrt{\frac{200.0 \text{ N}}{(1.25 \text{ m})(0.0548 \text{ kg})}} = \boxed{81.0 \text{ Hz}}$$

Fourth harmonic ($n = 4$):

$$f_1 = 2\sqrt{\frac{200.0 \text{ N}}{(1.25 \text{ m})(0.0548 \text{ kg})}} = \boxed{108 \text{ Hz}}$$

Part b)

Figure 13-5 Problem 65

First harmonic = purple ($n = 1$)

Second harmonic = green ($n = 2$)

Third harmonic = red ($n = 3$)

Fourth harmonic = blue ($n = 4$)

REFLECT

The frequencies of the higher harmonics are integer multiples of the fundamental frequency, $f_n = nf_1$.

13.69

SET UP

A steel string ($\rho_{\text{steel}} = 7800 \text{ kg/m}^3$) has an unstretched length of $L = 0.325$ m and a diameter of $d = 0.25 \times 10^{-3}$ m. By setting the two expressions for the speed of a wave on a string, $v_p = \sqrt{\frac{F}{\mu}}$ and $v_p = \lambda f$, equal to one another, we can solve for the tension in the string. We can rewrite the linear mass density in terms of the density of steel, the volume of the string, and L. The fundamental frequency of this string is $f_1 = 660$ Hz.

SOLVE

$$L = \frac{n\lambda}{2}$$

$$\lambda = \frac{2L}{n}$$

$$\lambda_1 = 2L$$

$$v_p = \sqrt{\frac{F}{\mu}} = \sqrt{\frac{F}{\left(\frac{M}{L}\right)}} = \sqrt{\frac{FL}{\rho V}} = \sqrt{\frac{FL}{\rho\left(\pi\left(\frac{d}{2}\right)^2 L\right)}} = \sqrt{\frac{4F}{\rho \pi d^2}} = \lambda_1 f_1 = (2L)f_1$$

$$F = \pi L^2 f^2 \rho d^2 = \pi (0.325 \text{ m})^2 (660 \text{ Hz})^2 \left(7800\frac{\text{kg}}{\text{m}^3}\right)(0.25 \times 10^{-3} \text{ m})^2 = \boxed{70.5 \text{ N}}$$

REFLECT
The Young's modulus of steel is 200×10^9 Pa. The strain due to the tensile stress on the string is 0.7%, so we can safely assume that the stretched length of the string is approximately equal to its unstretched length.

13.71

SET UP
Two successive harmonics of an organ pipe are 228.6 Hz and 274.3 Hz. We are asked to determine whether the pipe is open on both ends or open on one end and closed at the other end. Taking the ratio of these two frequencies will tell us the values of n to which these frequencies correspond. Only odd harmonics are allowed in a closed pipe, which means the integers in this ratio must differ by 2. An open pipe has harmonics with successive values of n.

SOLVE
Closed:

$$\frac{f_{n+2}}{f_n} = \frac{(n+2)f_1}{nf_1} = \frac{n+2}{n}$$

$$\frac{274.3 \text{ Hz}}{228.6 \text{ Hz}} = \frac{6}{5} \stackrel{?}{=} \frac{n+2}{n}$$

The integers in this fraction must both be odd if the pipe is open on one end and closed on the other end. Therefore, it is $\boxed{\text{not a closed pipe}}$.

Open:

$$\frac{f_{n+1}}{f_n} = \frac{(n+1)f_1}{nf_1} = \frac{n+1}{n}$$

$$\frac{274.3 \text{ Hz}}{228.6 \text{ Hz}} = \frac{6}{5} \stackrel{?}{=} \frac{n+1}{n}$$

The integers in this fraction must be successive if the pipe is open on both ends, which 5 and 6 are. Therefore, it is $\boxed{\text{an open pipe}}$.

REFLECT

We didn't have to calculate fully for both cases. Once we divided the frequencies and saw that the fraction was 6/5, we knew right away that it had to be an open pipe.

13.77

SET UP

The longest pipe in a pipe organ has a length of 9.75 m and the shortest is 0.0191 m. Assuming the speed of air is 343 m/s, we can find the range of the organ by calculating the fundamental frequencies of these pipes.

SOLVE

Fundamental frequencies for open pipes:

$$f_n = n\left(\frac{v_{\text{sound}}}{2L}\right)$$

$$f_{1,\text{low}} = \frac{v_{\text{sound}}}{2L} = \frac{\left(343\frac{m}{s}\right)}{2(9.75 \text{ m})} = \boxed{17.6 \text{ Hz}}$$

$$f_{1,\text{high}} = \frac{v_{\text{sound}}}{2L} = \frac{\left(343\frac{m}{s}\right)}{2(0.0191 \text{ m})} = \boxed{8.98 \times 10^3 \text{ Hz}}$$

Fundamental frequencies for closed pipes:

$$f_n = n\left(\frac{v_{\text{sound}}}{4L}\right)$$

$$f_{1,\text{low}} = \frac{v_{\text{sound}}}{4L} = \frac{\left(343\frac{m}{s}\right)}{4(9.75 \text{ m})} = \boxed{8.79 \text{ Hz}}$$

$$f_{1,\text{high}} = \frac{v_{\text{sound}}}{4L} = \frac{\left(343\frac{m}{s}\right)}{4(0.0191 \text{ m})} = \boxed{4.49 \times 10^3 \text{ Hz}}$$

REFLECT

The largest range of frequencies occurs when the long pipe acts as a closed pipe and the short pipe acts as an open one.

13.81

SET UP

A guitar string under 100.0 N of tension is supposed to have a frequency of 110.0 Hz. But, when played at the same time as a reference tone at 110.0 Hz, beats at 2.0 Hz are heard. The tension in the string is decreased, and the beat frequency increases, which means the string is becoming more out of tune. For a guitar, the length of the string and, thus, the wavelength are fixed. Therefore, the frequency is proportional to the square root of the tension. When the tension is decreased, the frequency of the sound will also decrease. Because the beat frequency increases when the tension in the string decreases, the initial frequency of the guitar string must be below the desired frequency. Once we know the initial frequency of the guitar, we can set up a ratio between the tensions and the frequencies and solve for the final desired tension when the string is in tune.

SOLVE
Initial frequency:

$$f_{beats} = |f_2 - f_1|$$

$$f_2 = f_1 \pm f_{beats} = (110.0 \text{ Hz}) \pm (2.0 \text{ Hz}) = 112.0 \text{ Hz or } 108.0 \text{ Hz}$$

Because the beat frequency increases when the tension in the string decreases, the initial frequency of the guitar string must be below the desired frequency, or $f_1 = 108.0$ Hz.

Tension:

$$\frac{v_1}{v_2} = \frac{\lambda f_1}{\lambda f_2} = \frac{\sqrt{\frac{F_1}{\mu}}}{\sqrt{\frac{F_2}{\mu}}}$$

$$\frac{f_1}{f_2} = \sqrt{\frac{F_1}{F_2}}$$

$$F_2 = F_1 \left(\frac{f_2}{f_1}\right)^2 = (100.0 \text{ N})\left(\frac{110.0 \text{ Hz}}{108.0 \text{ Hz}}\right)^2 = \boxed{103.7 \text{ N}}$$

REFLECT
We are told that decreasing the tension increases the beat frequency. Therefore, the tension should be higher than 100 N when the guitar is in tune.

Get Help: P'Cast 13.8 – Guitar Tuning

13.85

SET UP
Rush hour traffic lasts for 4 h each day. A nearby resident plans to convert this sound energy into a more useful form of energy by collecting the sound with a 1-m² microphone that has an efficiency of 30%. The sound level of the traffic at the location of the microphone is $\beta = 100$ dB. The definition of the sound level relates the intensity of the sound at a given position to the reference intensity, 10^{-12} W/m². The total power incident the microphone is equal to the intensity at the microphone multiplied by the surface area of the microphone. The total sound energy due to the traffic is equal to the power multiplied by the time interval the traffic is active, or 4 h. Because the microphone only absorbs 30% of the total energy that hits it, we need to multiply the total energy by 0.30 to find the total energy collected by the microphone.

SOLVE
Power:

$$\beta = (10 \text{ dB}) \log\left(\frac{I}{10^{-12}\frac{\text{W}}{\text{m}^2}}\right)$$

$$I = \left(10^{-12}\frac{\text{W}}{\text{m}^2}\right) 10^{\frac{\beta}{(10 \text{ dB})}} = \left(10^{-12}\frac{\text{W}}{\text{m}^2}\right) 10^{\frac{(100 \text{ dB})}{(10 \text{ dB})}} = 10^{-2}\frac{\text{W}}{\text{m}^2}$$

$$I = \frac{P}{A}$$

$$P = IA = \left(10^{-2}\frac{\text{W}}{\text{m}^2}\right)(1 \text{ m}^2) = 10^{-2} \text{ W}$$

Energy collected:

$$P = \frac{E_{\text{total}}}{t}$$

$$E_{\text{total}} = Pt = (10^{-2} \text{ W})\left(4 \text{ h} \times \frac{3600 \text{ s}}{1 \text{ h}}\right) = 144 \text{ J}$$

$$E_{\text{collected}} = (0.30) E_{\text{total}} = (0.30)(144 \text{ J}) = \boxed{40 \text{ J}}$$

She would do better by completely replacing the plan with something else. Barring that, she could move the collector closer to the highway (above the sound-blocking barrier would be best) or make it much larger (much of the cost would be in electronics, not the collector). Focusing wouldn't do much in the horizontal direction as the source is diffuse, but vertical focusing reflectors could help.

REFLECT
The 40 J is not very much energy. If this amount of energy were in the form of heat, it would increase the temperature of 1 kg of water by only 0.01°C.

13.89

SET UP
We can use the formula we derived in Problem 13.88 to relate an increase in sound level to an increase in intensity.

SOLVE
Part a)

$$\Delta\beta = (10 \text{ dB}) \log\left(\frac{I_2}{I_1}\right) = 1 \text{ dB}$$

$$I_2 = I_1 10^{\frac{1}{(10 \text{ dB})}} = \boxed{1.26 I_1}$$

Part b)

$$\Delta\beta = (10 \text{ dB})\log\left(\frac{I_2}{I_1}\right) = 20 \text{ dB}$$

$$I_2 = I_1 10^2 = \boxed{100 I_1}$$

REFLECT
We would expect the intensity to increase if there is an increase in the sound level.

13.95

SET UP
A bicyclist is moving toward a sheer wall at an unknown speed v while holding a tuning fork ringing at $f_{\text{emitted}} = 484$ Hz. The bicyclist detects a beat frequency of 6 Hz between the sound from the tuning fork and the sound reflecting off the wall. Because the bicycle is moving toward the wall, the reflected frequency will be higher than f_0. Once we know the reflected frequency, we can use $f'' = \left(\dfrac{v_{\text{sound}} + v}{v_{\text{sound}} - v}\right) f_{\text{emitted}}$ to calculate the speed of the bicycle. We'll assume the speed of sound in air to be 343 m/s.

SOLVE
Observed reflected frequency:
The bicycle is moving toward the wall, so we expect the observed reflected frequency to be higher than the original frequency. Therefore:

$$f_{\text{beats}} = |f_2 - f_1|$$
$$f_{\text{beats}} = |f'' - f_{\text{emitted}}|$$
$$f'' = f_{\text{beats}} + f_{\text{emitted}} = (484 \text{ Hz}) + (6 \text{ Hz}) = 490 \text{ Hz}$$

Speed of the bicycle:

$$f'' = \left(\frac{v_{\text{sound}} + v}{v_{\text{sound}} - v}\right) f_{\text{emitted}}$$

$$\frac{f''}{f_{\text{emitted}}}(v_{\text{sound}} - v) = v_{\text{sound}} + v$$

$$\left(\frac{f''}{f_{\text{emitted}}} - 1\right) v_{\text{sound}} = \left(\frac{f''}{f_{\text{emitted}}} + 1\right) v$$

$$v = \frac{\left(\dfrac{f''}{f_{\text{emitted}}} - 1\right)}{\left(\dfrac{f''}{f_{\text{emitted}}} + 1\right)} v_{\text{sound}} = \frac{\left(\dfrac{490 \text{ Hz}}{484 \text{ Hz}} - 1\right)}{\left(\dfrac{490 \text{ Hz}}{484 \text{ Hz}} + 1\right)} \left(343 \frac{\text{m}}{\text{s}}\right) = \boxed{2.11 \frac{\text{m}}{\text{s}}}$$

REFLECT
A speed of 2.11 m/s is a little under 5 mph, which is a reasonable speed for a person cycling while trying to hold a tuning fork.

13.97

SET UP

A bat emits a high-pitched sound at $f = 5.000 \times 10^4$ Hz while traveling at 10.0 m/s toward an insect that is traveling away from it. The bat observes the reflected wave that echoes off the insect at a frequency of $f_{\text{listener}} = 5.005 \times 10^4$ Hz. The general Doppler shift is given by $f_{\text{listener}} = \left(\dfrac{v_{\text{sound}} \pm v_{\text{listener}}}{v_{\text{sound}} \mp v_{\text{source}}} \right) f$, where the top signs correspond to motion "toward" and the bottom signs correspond to motion "away." The bat acts as a moving source traveling toward the observer at $v_{\text{source}} = 10.0$ m/s. The insect acts as a moving observer traveling away from the source at v_{listener}. This means we'll use the top sign $(-)$ for the source term and the bottom sign for the observer term $(-)$.

SOLVE

$$f_{\text{listener}} = \left(\frac{v_{\text{sound}} - v_{\text{listener}}}{v_{\text{sound}} - v_{\text{source}}} \right) f$$

$$v_{\text{listener}} = v_{\text{sound}} - \frac{f_{\text{listener}}}{f}(v_{\text{sound}} - v_{\text{source}})$$

$$= \left(343 \frac{\text{m}}{\text{s}} \right) - \frac{5.005 \times 10^4 \text{ Hz}}{5.000 \times 10^4 \text{ Hz}} \left(\left(343 \frac{\text{m}}{\text{s}} \right) - \left(10.0 \frac{\text{m}}{\text{s}} \right) \right) = \boxed{9.67 \frac{\text{m}}{\text{s}}}$$

REFLECT

If the mosquito doesn't change its path, the bat traveling at 10 m/s in the same direction will catch up to it and assuredly eat it.

General Problems

13.101

SET UP

The speed of sound in helium and air at 0°C is $v_{\text{He}} = 972$ m/s and $v_{\text{air}} = 331$ m/s, respectively. We can use the speed of sound and density of helium ($\rho_{\text{He}} = 0.1786$ kg/m³) to find the bulk modulus of helium. A person producing a frequency of 0.500 kHz in air will produce a different frequency if his respiratory tract is filled with helium. Since the effective length of the resonator remains constant, the wavelengths of the standing waves will also be constant, regardless of the gas in the resonator. Therefore, the ratio of the speeds of sound in helium and air will equal the ratio of the frequencies in helium and air. The addition of helium to the respiratory tract will affect the mix of frequencies excited in the resonator and, thus, the sound produced.

SOLVE

Part a)

$$v_{\text{He}} = \sqrt{\frac{B_{\text{He}}}{\rho_{\text{He}}}}$$

$$B_{\text{He}} = \rho_{\text{He}} v_{\text{He}}^2 = \left(0.1786 \frac{\text{kg}}{\text{m}^3} \right) \left(972 \frac{\text{m}}{\text{s}} \right)^2 = \boxed{1.69 \times 10^5 \text{ Pa}}$$

Part b)

$$\frac{v_{He}}{v_{air}} = \frac{\lambda f_{He}}{\lambda f_{air}}$$

$$f_{He} = \left(\frac{v_{He}}{v_{air}}\right) f_{air} = \frac{\left(972 \frac{m}{s}\right)}{\left(331 \frac{m}{s}\right)} (0.500 \text{ kHz}) = \boxed{1.47 \text{ kHz}}$$

Part c) The parts of the voice resonator that are in the air all have a frequency increase from the substitution of some helium into the air; the parts of the resonator that are in the body do not. This produces a different mix of frequencies and changes the character of the sound.

REFLECT
The speed of sound in sulfur hexafluoride (SF_6) at room temperature is about $0.44 v_{air}$. Introducing it into the respiratory tract will have the opposite effect of helium on the person's voice by making it deeper.

Get Help: Interactive Exercise – Tension
Interactive Exercise – Slinky
P'Cast 13.4 – Propagation Speeds

13.103

SET UP
The frequency of a musical note that is exactly an octave higher than another note is twice that of the lower note; correspondingly, the frequency of a note that is lower by an octave is half as big. A note that is two octaves above concert A (440 Hz) will, therefore, be four times higher. The ratio of the frequency of middle C to a note one octave below it will be equal to the square root of the ratio of the tensions in the string while playing those notes.

SOLVE
Part a)

One octave:

$$A' = 2A$$

Two octaves:

$$A'' = 2A' = 2(2A) = 4A = 4(440 \text{ Hz}) = \boxed{1760 \text{ Hz}}$$

Part b)

$$v_p = f\lambda$$

$$\frac{f_{\text{middle C}}}{f_{\text{octave below}}} = \frac{\left(\frac{v_{\text{middle C}}}{\lambda}\right)}{\left(\frac{v_{\text{octave below}}}{\lambda}\right)} = \frac{\left(\sqrt{\frac{F_{\text{middle C}}}{\mu}}\right)}{\left(\sqrt{\frac{F_{\text{octave below}}}{\mu}}\right)} = \sqrt{\frac{F_{\text{middle C}}}{F_{\text{octave below}}}}$$

$$F_{\text{octave below}} = (F_{\text{middle C}}) \left(\frac{f_{\text{octave below}}}{f_{\text{middle C}}}\right)^2 = (T) \left(\frac{f_{\text{octave below}}}{2 f_{\text{octave below}}}\right)^2 = \boxed{\frac{T}{4}}$$

REFLECT

We can safely assume the length of the vibrating string on the viola does not change, which means the wavelengths of the standing wave will not change. The frequency of the note corresponds to the pitch that we hear.

13.109

SET UP

A sound level of 120 dB was recorded from 10-Hz sound waves emitted from a volcanic eruption. The intensity of the sound at the location of the detector is related both to the sound level $\left(\beta = (10\text{ dB})\log\left(\dfrac{I}{10^{-12}\dfrac{\text{W}}{\text{m}^2}}\right)\right)$ and to the maximum displacement of the air s_{max} ($I = 2\pi^2 \rho v_p f^2 s_{max}^2$), where $\rho = 1.2\text{ kg/m}^3$. We can also relate the intensity to the amount of energy the wave would deliver to a 2.0 m × 3.0 m wall located at the site of the detector within 1.0 min, $I = \dfrac{P}{A} = \dfrac{E}{At}$; recall that power is energy transferred per time.

SOLVE

Part a)

Intensity:

$$\beta = (10\text{ dB})\log\left(\dfrac{I}{10^{-12}\dfrac{\text{W}}{\text{m}^2}}\right)$$

$$I = \left(10^{\frac{\beta}{(10\text{ dB})}}\right)\left(10^{-12}\dfrac{\text{W}}{\text{m}^2}\right)$$

Maximum displacement:

$$I = 2\pi^2 \rho v_p f^2 s_{max}^2 = \left(10^{\frac{\beta}{(10\text{ dB})}}\right)\left(10^{-12}\dfrac{\text{W}}{\text{m}^2}\right)$$

$$s_{max} = \sqrt{\dfrac{\left(10^{\frac{\beta}{(10\text{ dB})}}\right)\left(10^{-12}\dfrac{\text{W}}{\text{m}^2}\right)}{2\pi^2 \rho v_p f^2}} = \sqrt{\dfrac{\left(10^{\frac{(120\text{ dB})}{(10\text{ dB})}}\right)\left(10^{-12}\dfrac{\text{W}}{\text{m}^2}\right)}{2\pi^2\left(1.2\dfrac{\text{kg}}{\text{m}^3}\right)\left(343\dfrac{\text{m}}{\text{s}}\right)(10\text{ Hz})^2}} = \boxed{0.001\text{ m} = 1\text{ mm}}$$

Part b)

$$I = \dfrac{P}{A} = \dfrac{E}{At} = \left(10^{\frac{\beta}{10}}\right)\left(10^{-12}\dfrac{\text{W}}{\text{m}^2}\right)$$

$$E = \left(10^{\frac{\beta}{(10\text{ dB})}}\right)\left(10^{-12}\dfrac{\text{W}}{\text{m}^2}\right)At = \left(10^{\frac{(120\text{ dB})}{(10\text{ dB})}}\right)\left(10^{-12}\dfrac{\text{W}}{\text{m}^2}\right)((2.0\text{ m})(3.0\text{ m}))(60\text{ s}) = \boxed{4 \times 10^2\text{ J}}$$

REFLECT
Since we don't know the distance between the volcano and the sound detector, we have to assume the wall is located at the same distance away.

13.113

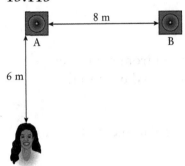

Figure 13-6 Problem 113

SET UP
Two 40-W speakers, labeled A and B, are playing the same 857.5-Hz tone in phase. The speakers are 8 m apart, and speaker A is 6 m in front of you. In order to determine the intensity where you are, we need to determine whether or not there is constructive interference by calculating the path difference between the sound waves from A and B and compare it to the wavelength. If there is fully constructive interference, the total intensity is the sum of the intensity of each speaker at that position, where the intensity of each speaker is $I = \dfrac{P}{4\pi r^2}$. The sound level is equal to $\beta = (10 \text{ dB}) \log \left(\dfrac{I_{\text{total}}}{10^{-12} \frac{\text{W}}{\text{m}^2}} \right)$.

SOLVE
Wavelength:

$$v_p = f\lambda$$

$$\lambda = \dfrac{v_p}{f} = \dfrac{\left(343\dfrac{\text{m}}{\text{s}}\right)}{857.5 \text{ Hz}} = 0.4 \text{ m}$$

Interference:

$$\Delta_{\text{pl}} = D_B - D_A = (\sqrt{(8 \text{ m})^2 + (6 \text{ m})^2}) - (6 \text{ m}) = (10 \text{ m}) - (6 \text{ m}) = 4 \text{ m}$$

$$\dfrac{\Delta_{\text{pl}}}{\lambda} = \dfrac{4 \text{ m}}{0.4 \text{ m}} = 10$$

which means there will be fully constructive interference.

Total Intensity:
$$I_{\text{total}} = I_A + I_B = \frac{P}{4\pi r_A^2} + \frac{P}{4\pi r_B^2} = \frac{P}{4\pi}\left(\frac{1}{r_A^2} + \frac{1}{r_B^2}\right) = \frac{40\text{ W}}{4\pi}\left(\frac{1}{(6\text{ m})^2} + \frac{1}{(10\text{ m})^2}\right) = \boxed{0.12\frac{\text{W}}{\text{m}^2}}$$

Sound level:
$$\beta = (10\text{ dB})\log\left(\frac{I_{\text{total}}}{10^{-12}\frac{\text{W}}{\text{m}^2}}\right) = (10\text{ dB})\log\left(\frac{\left(0.12\frac{\text{W}}{\text{m}^2}\right)}{\left(10^{-12}\frac{\text{W}}{\text{m}^2}\right)}\right) = \boxed{1\times 10^2\text{ dB}}$$

REFLECT

A sound level of 1×10^2 dB is equivalent to a car horn at a distance of 1 m.

Chapter 14
Thermodynamics I

Conceptual Questions

14.3 The warmer the day, the faster the molecules will move, so the sooner the smell will spread.

Get Help: Interactive Exercise – Syringe

14.9 Part a) They are the same temperature.

Part b) The steam can cause a more severe burn because of its latent heat of vaporization.

14.17 The two metals must be chosen to have different thermal expansion coefficients. Because the metals are attached directly, thermal expansion will bend the bimetallic strip, which in turn pushes or pulls on the switch.

Multiple-Choice Questions

14.27 B (reduced to one-quarter its original value).

$$v_{rms} = \sqrt{\frac{3kT}{m}}$$

$$v^2 \propto T$$

$$\frac{v_{rms,2}}{v_{rms,1}} = \frac{1}{2} = \sqrt{\frac{T_2}{T_1}}$$

$$T_2 = \frac{1}{4}T_1$$

Get Help: Interactive Exercise – Syringe
P'Cast 14.3 – Oxygen at Room

14.31 D (the temperature of each object will be the same). The two objects will eventually reach thermal equilibrium.

Get Help: Picture It – Temperature and Boiling Points

14.35 C (halved). The rate of heat transfer is inversely proportional to the thickness of the wall L.

$$H = k\frac{A}{L}(T_H - T_C)$$

$$\frac{H_2}{H_1} = \frac{\left(\frac{1}{L_2}\right)}{\left(\frac{1}{L_1}\right)} = \frac{L_1}{L_2} = \frac{L_1}{(2L_1)} = \frac{1}{2}$$

Get Help: Picture It – Thermal Current

Estimation/Numerical Analysis

14.41 After about 14 min, your muscles will shut down, and your heart will stop after about 28 min. Assume a body mass of $m_{body} = 70$ kg, a surface area of $A = 2$ m², and a thickness of fat through which the heat must flow of $L = 0.01$ m. Also assume that a temperature change of about 5°C (5 K) will shut down the muscles and a temperature change of about 10°C (10 K) will stop the heart. Further assume the water temperature just under the ice is about 1°C, so that $\Delta T_{body/water} = 37°C - 1°C = 36°C$, or 36 K. The specific heat of the body is about 3470 J/kg·K, and the thermal conductivity of human fat is $0.2 \frac{W}{m \cdot k}$.

Heat loss and time for the muscles to shut down:

$$Q = mc\Delta T$$

$$Q_{5K} = m_{body} c_{body} \Delta T_{5K} = (70 \text{ kg})\left(3470 \frac{J}{kg \cdot K}\right)(5 \text{ K}) \approx 1{,}214{,}500 \text{ J}$$

$$\frac{Q}{\Delta t} = k\frac{A}{L}(T_H - T_C)$$

$$\Delta t_{muscles} = \frac{Q_{5K} L}{kA \Delta T_{body/water}} = \frac{(1{,}214{,}500 \text{ J})(0.01 \text{ m})}{\left(0.2 \frac{W}{m \cdot K}\right)(2 \text{ m}^2)(36 \text{ K})} \approx 840 \text{ s} \times \frac{1 \text{ min}}{60 \text{ s}} \approx 14 \text{ min}$$

For the heart to stop, $\Delta T_{10K} = 2\Delta T_{5K}$, so that $Q_{10K} = 2Q_{5K}$. The time for this to occur, then, is about twice as long, or about 28 minutes.

14.47 About 2 times more heat is lost to radiation compared to conduction. Assume a body temperature of 37°C (310 K), an air and snow temperature of 0°C (273 K), and a body fat layer of thickness $d = 0.015$ m.

$$P = e\sigma A T^4$$

$$\frac{Q_{radiation}}{\Delta t} = e_{body} \sigma A_{body} T^4_{body}$$

$$Q_{radiation} = \Delta t \, e_{body} A_{body} \sigma T^4_{body}$$

$$= \Delta t A_{body} (0.97)\left(5.67 \times 10^{-8} \frac{W}{m^2 \cdot K^4}\right)(310 \text{ K})^4 \approx 507.927 \Delta t A_{body}$$

$$\frac{Q_{conduction}}{\Delta t} = k\frac{A}{L}(T_H - T_C)$$

$$Q_{conduction,total} = Q_{conduction,snow} + Q_{conduction,air}$$

$$Q_{conduction,total} = \Delta t\, k_{snow}\frac{A_{body}}{L}(T_{body} - T_{snow}) + \Delta t\, k_{air}\frac{A_{body}}{L}(T_{body} - T_{air})$$

$$Q_{conduction,total} = \Delta t\, k_{snow}\frac{A_{body}}{2d}(T_{body} - T_{snow}) + \Delta t\, k_{air}\frac{A_{body}}{2d}(T_{body} - T_{air})$$

$$Q_{conduction,total} = \Delta t A_{body}\frac{\left(0.16\frac{W}{m\cdot K}\right)(310\,K - 273\,K)}{2(0.015\,m)} + \Delta t A\frac{\left(0.024\frac{W}{m\cdot K}\right)(310\,K - 273\,K)}{2(0.015\,m)}$$

$$Q_{conduction,total} = 226.933\, \Delta t A_{body}$$

$$\frac{Q_{radiation}}{Q_{conduction,total}} = \frac{507.927\, \Delta t A_{body}}{226.933\, \Delta t A_{body}} \approx 2.2$$

Get Help: Picture It – Thermal Current

Problems

14.57

SET UP

To see that there exists a numerical value that is the same on both the Celsius and Fahrenheit scales, consider thermal energy being added or removed from an ideal gas. The average kinetic energy per molecule of the ideal gas is the same regardless of the temperature scale used, but the degree units on the Celsius and Fahrenheit scales are different. While a plot of average kinetic energy per molecule in the gas versus temperature yields a straight line using either scale, the lines have a different slope when plotted using Fahrenheit temperatures or Celsius temperatures. Therefore, these lines must cross. At the crossing point, both temperature scales have the same numerical value, and both describe the same system and therefore the same temperature.

SOLVE

Special value where $T_C = T_F$:

$$T_C = \frac{5}{9}(T_F - 32) = \frac{5}{9}(T_C - 32)$$

$$\frac{4}{9}T_C = (-32)\left(\frac{5}{9}\right)$$

$$\boxed{T_C = -40 = T_F}$$

REFLECT

The Celsius and Kelvin scales do not share a numerical value.

Get Help: Picture It – Temperature and Boiling Points
P'Cast 14.1 – Cold, Hot, and in Between

14.61

SET UP

An unknown gas has a volume $V = 4.13$ L and pressure $p = 10.0$ atm at a temperature $T = 293$ K. We can use the ideal gas law, with $R = 0.0821 \frac{\text{L} \cdot \text{atm}}{\text{mol} \cdot \text{K}}$, to find the number of moles of gas. The gas sample has a mass of 55.0 g; dividing this by the number of moles yields the molar mass of the unknown gas. We can use the molar mass and a periodic table to determine the identity of the unknown gas.

SOLVE
Number of moles:

$$pV = nRT$$

$$n = \frac{pV}{RT} = \frac{(10.0 \text{ atm})(4.13 \text{ L})}{\left(0.0821 \frac{\text{L} \cdot \text{atm}}{\text{mol} \cdot \text{K}}\right)(293 \text{ K})} = 1.717 \text{ mol}$$

Molar mass:

$$\frac{55.0 \text{ g}}{1.717 \text{ mol}} = 32.0 \frac{\text{g}}{\text{mol}}$$

This is the molar mass of $\boxed{O_2}$.

REFLECT
It is important to use consistent units when performing the calculation. The pressure and volume are given in non-SI units, so it makes sense to use $R = 0.0821 \frac{\text{L} \cdot \text{atm}}{\text{mol} \cdot \text{K}}$.

Get Help: Interactive Exercise – Syringe

14.65

SET UP

Water vapor (H_2O, molecular mass = 18.0 u) and oxygen (O_2, molecular mass = 32.0 u) are both present in the atmosphere. Since the gases are allowed to travel three in dimensions, the average kinetic energy of a single molecule of the gas is $K_{\text{translational,average}} = \frac{1}{2}m(v^2)_{\text{average}} = \frac{3}{2}kT$, and the square root of $(v^2)_{\text{average}}$ is v_{rms}. We assume the temperature is constant.

SOLVE
Root-mean-square speed:

$$K_{\text{translational,average}} = \frac{1}{2}m(v^2)_{\text{average}} = \frac{3}{2}kT$$

$$v_{\text{rms}} = \sqrt{(v^2)_{\text{average}}} = \sqrt{\frac{3kT}{m}}$$

154 Chapter 14 Thermodynamics I

Ratio:

$$\frac{v_{\text{rms,water}}}{v_{\text{rms,oxygen}}} = \frac{\left(\sqrt{\frac{3kT}{m_{\text{water}}}}\right)}{\left(\sqrt{\frac{3kT}{m_{\text{oxygen}}}}\right)} = \sqrt{\frac{m_{\text{oxygen}}}{m_{\text{water}}}} = \sqrt{\frac{32.0 \text{ u}}{18.0 \text{ u}}} = \sqrt{\frac{16}{9}} = \frac{4}{3}$$

REFLECT
The lighter molecule (water) has a higher v_{rms} at a given temperature, which makes sense.

Get Help: Interactive Exercise – Syringe
P'Cast 14.3 – Oxygen at Room

14.67

SET UP
The mean free path for O_2 molecules at a temperature of 300 K and a pressure of 1.00 atm is 7.10×10^{-8} m. We can rearrange the expression for the mean free path, $\lambda = \frac{kT}{4\sqrt{2}\pi r^2 p}$, in order to solve for the radius of an oxygen molecule.

SOLVE

$$\lambda = \frac{kT}{4\sqrt{2}\pi r^2 p}$$

$$r = \sqrt{\frac{kT}{4\sqrt{2}\pi \lambda p}} = \sqrt{\frac{\left(1.381 \times 10^{-23} \frac{\text{J}}{\text{K}}\right)(300 \text{ K})}{4\sqrt{2}\pi(7.10 \times 10^{-8} \text{ m})\left(1.00 \text{ atm} \times \frac{1.01 \times 10^5 \text{ Pa}}{1 \text{ atm}}\right)}} = \boxed{2 \times 10^{-10} \text{ m}}$$

REFLECT
The radius of an oxygen atom is 0.6×10^{-10} m, and the length of an oxygen–oxygen double bond is 1.21×10^{-10} m. If we model O_2 as two solid spheres 0.6×10^{-10} m in radius separated by 1.21×10^{-10} m, we will get a diameter of 3.61×10^{-10} m, which agrees well with our answer.

Get Help: Interactive Exercise – Syringe

14.71

SET UP
The temperature of a silver pin that is 5.00 cm long decreases by 152 K. The coefficient of linear expansion of silver is 19.5×10^{-6} K^{-1}. The final length of the pin after cooling is equal to the initial length L_0 plus the change in the length due to the temperature change, $\Delta L = \alpha L_0 \Delta T$.

SOLVE

$$\Delta L = \alpha L_0 \Delta T$$

$$L_f = L_0 + \Delta L = L_0 + (\alpha L_0 \Delta T) = L_0(1 + \alpha \Delta T)$$

$$= (5.00 \text{ cm})(1 + (19.5 \times 10^{-6} \text{ K}^{-1})(-152 \text{ K})) = \boxed{4.99 \text{ cm}}$$

REFLECT

We expect the pin to get smaller as it cools down, so $L_f < L_0$.

Get Help: Picture It – Linear Expansion
P'Cast 14.4 – An Expanding Bridge

14.75

SET UP

A cylinder of solid aluminum has an initial length $L_0 = 5.00$ m and an initial radius $R_0 = 2.00$ cm. The temperature of the aluminum is increased by 25.0 K. The length and radius of the cylinder both increase independently due to the temperature increase according to $\Delta L = \alpha L_0 \Delta T$, where the coefficient of linear expansion of aluminum is 22.2×10^{-6} K^{-1}. Since the mass of the cylinder remains constant but the volume increases, the density of the aluminum cylinder will decrease. We can set up a ratio of the densities in order to calculate the factor by which the density decreases and the volume increases.

SOLVE

Part a)

$$\Delta L = \alpha L_0 \Delta T = (22.2 \times 10^{-6} \text{ K}^{-1})(5.00 \text{ m})(25.0 \text{ K}) = \boxed{0.00278 \text{ m}}$$

Part b)

Final radius:

$$\Delta R = \alpha R_0 \Delta T$$

$$R_f = R_0 + \Delta R = R_0 + (\alpha R_0 \Delta T) = R_0(1 + \alpha \Delta T)$$

$$= (2.00 \text{ cm})(1 + (22.2 \times 10^{-6} \text{ K}^{-1})(25.0 \text{ K})) = 2.00111 \text{ cm}$$

Change in density:

$$\frac{\rho_f}{\rho_0} = \frac{\left(\frac{m}{V_f}\right)}{\left(\frac{m}{V_0}\right)} = \frac{V_0}{V_f} = \frac{\pi R_0^2 L_0}{\pi R_f^2 L_f} = \frac{R_0^2 L_0}{R_f^2 L_f} = \frac{(2.00 \text{ cm})^2 (5.00 \text{ m})}{(2.00111 \text{ cm})^2 (5.00278 \text{ m})} = 0.998$$

The density $\boxed{\text{decreases by 0.2\%}}$ upon heating.

Part c)

$$\frac{V_f}{V_0} = \left(\frac{\rho_f}{\rho_0}\right)^{-1} = (0.998)^{-1} = 1.002$$

The volume $\boxed{\text{increases by 0.2\%}}$ upon heating.

REFLECT

The mass of the cylinder must remain constant since we are not adding or removing anything from it; all we are doing is increasing its size. We could have also used the approximate relationship for the thermal expansion in three dimensions: $V_f = V_0 + \Delta V = V_0(1 + 3\alpha \Delta T)$.

Get Help: Picture It – Linear Expansion
P'Cast 14.4 – An Expanding Bridge

14.79

SET UP

You want to heat up 0.25 kg of water from 20.0°C to 95.0°C. The minimum amount of heat required to accomplish that is given by $Q = mc\Delta T$, where c is the specific heat of water $c = 4186 \frac{J}{kg \cdot K}$.

SOLVE

$$Q = mc\Delta T = (0.25 \text{ kg})\left(4186 \frac{J}{kg \cdot K}\right)(75.0 \text{ K}) = \boxed{7.8 \times 10^4 \text{ J} = 78 \text{ kJ}}$$

REFLECT

A temperature difference of 75.0°C is the same as a temperature difference of 75.0 K. Accordingly, sometimes you'll see the specific heat of water as $c = 4186 \frac{J}{kg \cdot °C}$.

Get Help: Picture It – Calorimetry
P'Cast 14.5 – Camping Thermodynamics

14.83

SET UP

The temperature of a lake was increased from 283 K to 288 K when 1.70×10^{14} J was transferred to it. We can calculate the mass of the lake from $Q = mc\Delta T$ and the specific heat of the lake $\left(4186 \frac{J}{kg \cdot K}\right)$.

SOLVE

$$Q = mc\Delta T$$

$$m = \frac{Q}{c\Delta T} = \frac{1.70 \times 10^{14} \text{ J}}{\left(4186 \frac{J}{kg \cdot K}\right)(2.88 \text{ K} - 283 \text{ K})} = \boxed{8.12 \times 10^9 \text{ kg}}$$

REFLECT

The lake has a volume of approximately 8.12×10^9 L. For comparison, the volume of Lake Superior, the largest of the Great Lakes, is 1.2×10^{16} L.

Get Help: Picture It – Calorimetry
P'Cast 14.5 – Camping Thermodynamics

14.89

SET UP

Each stroke of a hacksaw supplies 30.0 J of heat to a 20.0-g steel bolt. We can calculate the total heat necessary to raise the temperature of the bolt by 60.0 K from $Q = mc\Delta T$, where $c = 452 \frac{J}{kg \cdot K}$. The total heat divided by the heat supplied per stroke will give the total number of strokes required.

SOLVE
Total heat required:

$$Q = mc\Delta T = (0.0200 \text{ kg})\left(452 \frac{\text{J}}{\text{kg} \cdot \text{K}}\right)(60.0 \text{ K}) = 542.4 \text{ J}$$

Number of strokes:

$$n_{\text{strokes}} = \frac{\text{Total heat}}{\text{Heat supplied by one stroke}} = \frac{542.4 \text{ J}}{30.0 \text{ J}} = 18.1 = \boxed{19 \text{ strokes}}$$

REFLECT
The 19th stroke will cause an overall temperature increase of more than 60.0 K, but 18 strokes will only increase the temperature by 59.7 K.

Get Help: Picture It – Calorimetry
P'Cast 14.5 – Camping Thermodynamics

14.93

SET UP
The total heat required to melt 0.400-kg of copper initially at 293 K is equal to the sum of the heat necessary to raise the temperature of the sample to the melting point ($T_f = 1357$ K) and the heat necessary to convert the entire solid into liquid. The heat necessary to increase the temperature of the object is given by $Q = mc\Delta T$, and the heat necessary to convert from solid to liquid is related to the latent heat of fusion, $Q = mL_F$. The specific heat of copper is $c = 387 \frac{\text{J}}{\text{kg} \cdot \text{K}}$, and the latent heat of fusion for copper in this problem is $L_F = 209 \times 10^3 \frac{\text{J}}{\text{kg}}$.

SOLVE
Increase temperature to melting point:

$$Q = mc\Delta T = (0.400 \text{ kg})\left(387 \frac{\text{J}}{\text{kg} \cdot \text{K}}\right)(1064 \text{ K}) = 164{,}707 \text{ J}$$

Melting the copper:

$$Q = mL_F = (0.400 \text{ kg})\left(209 \times 10^3 \frac{\text{J}}{\text{kg}}\right) = 83{,}600 \text{ J}$$

Total heat:

$$Q_{\text{total}} = (164{,}707 \text{ J}) + (83{,}600 \text{ J}) = 248{,}307 \text{ J} = \boxed{248 \text{ kJ}}$$

REFLECT
We have to increase the temperature of the copper by over 1000 K, so it makes sense that most of the heat added goes toward this process rather than melting.

14.97

SET UP
We want to see which sample of iron, copper, or water will melt the most from a huge block of ice. The iron sample has a mass of $m_{\text{Fe}} = 50.0$ g and an initial temperature of 120°C; the copper sample has a mass of $m_{\text{Cu}} = 60.0$ g and an initial temperature of 150°C; and the water

has a mass of $m_{water} = 30.0$ g and an initial temperature of 40.0°C. Assuming the sample and the ice form a thermally isolated system, the amount of heat leaving the sample is equal to the heat gained by the ice due to conservation of energy. The total heat lost by the sample is equal to the heat lost to lower its temperature from its initial value to 0°C. The heat gained by the ice goes into heating it from −5.00°C to 0°C and then melting the ice. The heat necessary to change the temperature of an object is given by $Q = mc\Delta T$, and the heat necessary to convert from liquid to solid is related to the latent heat of fusion by $Q = mL_F$. The specific heats of iron, copper, liquid water, and ice are $c_{Fe} = 452 \frac{J}{kg \cdot K}$, $c_{Cu} = 387 \frac{J}{kg \cdot K}$, $c_{water} = 4186 \frac{J}{kg \cdot K}$, and $c_{ice} = 2093 \frac{J}{kg \cdot K}$, respectively. The latent heat of fusion for water is $L_F = 334 \times 10^3 \frac{J}{kg}$. We can calculate the mass of ice melted in each case and compare these amounts to determine which sample melts the most.

SOLVE
Iron:

$$Q = mc\Delta T$$

$$Q = mL_F$$

$$Q_{ice,-5°C \text{ to } 0°C} + Q_{ice \text{ to water}} + Q_{Fe, 120°C \text{ to } 0°C} = 0$$

$$m_{ice}c_{ice}\Delta T_{-5°C \text{ to } 0°C} + m_{ice}L_F + m_{Fe}c_{Fe}\Delta T_{120°C \text{ to } 0°C} = 0$$

$$m_{ice} = \frac{-m_{Fe}c_{Fe}\Delta T_{120°C \text{ to } 0°C}}{c_{ice}\Delta T_{-5°C \text{ to } 0°C} + L_F} = \frac{-(50.0 \text{ g})\left(452\frac{J}{kg \cdot K}\right)(-120 \text{ K})}{\left(2093\frac{J}{kg \cdot K}\right)(5.00 \text{ K}) + \left(334 \times 10^3 \frac{J}{kg}\right)} = \boxed{7.9 \text{ g}}$$

Copper:

$$Q_{ice,-5°C \text{ to } 0°C} + Q_{ice \text{ to water}} + Q_{Cu, 150°C \text{ to } 0°C} = 0$$

$$m_{ice}c_{ice}\Delta T_{-5°C \text{ to } 0°C} + m_{ice}L_F + m_{Cu}c_{Cu}\Delta T_{150°C \text{ to } 0°C} = 0$$

$$m_{ice} = \frac{-m_{Cu}c_{Cu}\Delta T_{150°C \text{ to } 0°C}}{c_{ice}\Delta T_{-5°C \text{ to } 0°C} + L_F} = \frac{-(60.0 \text{ g})\left(387\frac{J}{kg \cdot K}\right)(-150 \text{ K})}{\left(2093\frac{J}{kg \cdot K}\right)(5.00 \text{ K}) + \left(334 \times 10^3 \frac{J}{kg}\right)} = \boxed{1.0 \times 10^1 \text{ g}}$$

Water:

$$Q_{ice,-5°C \text{ to } 0°C} + Q_{ice \text{ to water}} + Q_{water, 40°C \text{ to } 0°C} = 0$$

$$m_{ice}c_{ice}\Delta T_{-5°C \text{ to } 0°C} + m_{ice}L_F + m_{water}c_{water}\Delta T_{40°C \text{ to } 0°C} = 0$$

$$m_{ice} = \frac{-m_{water}c_{water}\Delta T_{40°C \text{ to } 0°C}}{c_{ice}\Delta T_{-5°C \text{ to } 0°C} + L_F} = \frac{-(30.0 \text{ g})\left(4186\frac{J}{kg \cdot K}\right)(-40.0 \text{ K})}{\left(2093\frac{J}{kg \cdot K}\right)(5.00 \text{ K}) + \left(334 \times 10^3 \frac{J}{kg}\right)} = \boxed{14.6 \text{ g}}$$

The $\boxed{\text{water will melt the most ice}}$.

REFLECT

Since we assume the ice block was really large, we do not have to worry about the sample and melted ice reaching an equilibrium temperature higher than 0°C. Even though the water had the smallest mass, it melted the most ice mainly due to water's large specific heat.

14.101

SET UP

A star radiates 1000 times more power than our Sun, but its temperature is only 70% of the Sun's temperature. The radiated power is proportional to both the surface area and the temperature to the fourth power: $P = e\sigma A T^4$. We can set up a ratio between the power radiated by each star in order to calculate the radius of the star in terms of the radius of the Sun, $R_{Sun} = 6.96 \times 10^8$ m. We assume that both stars are perfect emitters and that they are both perfectly spherical.

SOLVE

$$P = e\sigma A T^4$$

$$\frac{P_{star}}{P_{Sun}} = \frac{e\sigma A_{star} T^4_{star}}{e\sigma A_{star} T^4_{Sun}} = \frac{(4\pi R^2_{star})T^4_{star}}{(4\pi R^2_{Sun})T^4_{Sun}} = \frac{R^2_{star} T^4_{star}}{R^2_{Sun} T^4_{Sun}}$$

$$R_{star} = R_{Sun}\sqrt{\frac{P_{star} T^4_{Sun}}{P_{Sun} T^4_{star}}} = R_{Sun}\sqrt{\frac{(1000 P_{Sun})T^4_{Sun}}{P_{Sun}(0.7T_{Sun})^4}} = (6.96 \times 10^8 \text{ m})\sqrt{\frac{1000}{(0.7)^4}} = \boxed{4.49 \times 10^{10} \text{ m}}$$

REFLECT

This is about 65 times larger than the radius of the Sun. Changing the temperature has a much larger effect on the power than changing the surface area since it is raised to the fourth power.

14.103

SET UP

A glass window is 0.300 m × 1.50 m and 1.20×10^{-3} m thick. We can calculate the rate of heat transfer through the window by conduction by $H = k\frac{A}{L}(T_H - T_C)$. The temperature difference across the window is 17.0 K, and the thermal conductivity of window glass is $k = 0.96\frac{\text{W}}{\text{m} \cdot \text{K}}$.

SOLVE

$$H = k\frac{A}{L}(T_H - T_C) = \left(0.96\frac{\text{W}}{\text{m} \cdot \text{K}}\right)\frac{(0.300 \text{ m})(1.50 \text{ m})}{1.20 \times 10^{-3} \text{ m}}(17.0 \text{ K}) = \boxed{6100 \text{ W} = 6.1 \text{ kW}}$$

REFLECT

Unlike radiation, the rate of heat transfer by conduction depends on the temperature difference between two objects.

Get Help: Picture It – Thermal Current

General Problems

14.107

SET UP

A brick wall is composed of 19.0-cm-long bricks ($\alpha_{brick} = 5.5 \times 10^{-6}$ °C^{-1}) and 1.00-cm-long sections of mortar ($\alpha_{mortar} = 8.0 \times 10^{-6}$ °C^{-1}). A 20.0-m-long wall will be made up of 100 of these repeating brick + mortar units. In order to calculate the expansion on this wall due to a temperature increase of 25°C, we need to calculate the expansion of one brick and one piece of mortar due to this temperature increase. The total expansion of the wall will be equal to 100 times the expansion of one brick + mortar unit.

SOLVE

One brick + mortar unit:

$$\Delta L = \alpha L_0 \Delta T$$

$$\Delta L_{brick} = \alpha_{brick} L_{brick,0} \Delta T = (5.5 \times 10^{-6} \text{ °C}^{-1})(19.0 \text{ cm})(25\text{°C}) = 0.00261 \text{ cm}$$

$$\Delta L_{mortar} = \alpha_{mortar} L_{mortar,0} \Delta T = (8.0 \times 10^{-6} \text{ °C}^{-1})(1.00 \text{ cm})(25\text{°C}) = 0.000200 \text{ cm}$$

$$\Delta L_{total} = \Delta L_{brick} + \Delta L_{mortar} = (0.00261 \text{ cm}) + (0.000200 \text{ cm}) = 0.00281 \text{ cm}$$

There are 100 brick + mortar units in a 20-m-long wall, so:

$$\Delta L_{wall} = 100 \Delta L_{total} = 100(0.00281 \text{ cm}) = 0.28 \text{ cm}$$

The length of the wall increases by 0.28 cm.

REFLECT

A value of 2.8 mm seems to be a reasonable value for a wall to expand.

Get Help: Picture It – Linear Expansion
P'Cast 14.4 – An Expanding Bridge

14.111

SET UP

A person collects sound waves from traffic and converts them into an electrical signal used to heat 5 kg of water. After 7 days the temperature of the water increased 0.01 K. Using the specific heat of water $\left(c = 4186 \dfrac{\text{J}}{\text{kg} \cdot \text{K}}\right)$, we can calculate the energy (in the form of heat) required to cause this temperature increase. The total acoustic power "caught" by the transducer is equal to this energy divided by 7 days.

SOLVE

Heat required to raise the temperature:

$$Q = mc\Delta T = (5 \text{ kg})\left(4186 \dfrac{\text{J}}{\text{kg} \cdot \text{K}}\right)(0.01 \text{ K}) = 209.3 \text{ J}$$

Acoustic power collected:

$$P = \frac{\Delta E}{\Delta t} = \frac{209.3 \text{ J}}{\left(7 \text{ d} \times \frac{24 \text{ h}}{1 \text{ d}} \times \frac{3600 \text{ s}}{1 \text{ h}}\right)} = \boxed{3 \times 10^{-4} \text{ W}}$$

REFLECT

An efficiency of 100% means all of the acoustic energy is converted into electrical energy. In actuality, the efficiency would be (much) lower than 100%, so 3×10^{-4} W is the minimum power that the transducer collects.

14.113

SET UP

The total heat required to condense 5.00×10^{-3} kg of helium initially at 30.0°C is equal to the sum of the heat necessary to lower the temperature of the sample to the boiling point ($T_V = -268.93$°C) and the heat necessary to convert the entire gas into liquid. The heat transfer necessary to decrease the temperature of the helium is given by $Q = mc\Delta T$, and the heat necessary to convert from gas to liquid is related to the latent heat of vaporization by $Q = mL_V$. The specific heat of helium is $c = 5193\frac{\text{J}}{\text{kg} \cdot °\text{C}}$, and the latent heat of vaporization for helium is $L_V = 21{,}000\frac{\text{J}}{\text{kg}}$.

SOLVE

$$Q = mc\Delta T$$

$$Q = mL_V$$

$$Q_{\text{total}} = Q_{30°\text{C to} -268.93°\text{C}} + Q_{\text{gas to liquid}} = mc\Delta T_{30°\text{C to} -268.93°\text{C}} + mL_V = m(c\Delta T_{30°\text{C to} -268.93°\text{C}} + L_V)$$

$$= (5.00 \times 10^{-3} \text{ kg})\left(\left(5193\frac{\text{J}}{\text{kg} \cdot °\text{C}}\right)(-298.93°\text{C}) + \left(-21{,}000\frac{\text{J}}{\text{kg}}\right)\right) = \boxed{-7.87 \times 10^3 \text{ J}}$$

REFLECT

The majority of the heat lost by the helium goes into cooling the gas by almost 300°C.

14.117

SET UP

A man has a surface area of $A = 2.1$ m² and a skin temperature of 303 K. Normally 80% of the food calories (that is, kcal) he eats are converted into heat. The man is in a 293-K room and wants to know how many food calories x he needs to eat in a 24-h period in order to maintain his current body temperature. The net radiated energy is equal to the energy he radiates minus the energy the air radiates back into him. The power radiated is given by $P = e\sigma AT^4$, where we assume $e = 1$. The conversion between kcal and joules is 1 kcal = 4186 J.

SOLVE

$$P_{\text{net}} = P_{\text{body}} - P_{\text{air}}$$

$$\frac{\Delta E_{\text{net}}}{\Delta t} = P_{\text{body}} - P_{\text{air}}$$

$$\Delta E_{net} = (P_{body} - P_{air})\Delta t$$

$$(0.8x)\left(\frac{4186 \text{ J}}{1 \text{ kcal}}\right) = (e\sigma A T_{body}^4 - e\sigma A T_{air}^4)\Delta t = (T_{body}^4 - T_{air}^4)e\sigma \Delta t$$

$$x = \frac{(T_{body}^4 - T_{air}^4)e\sigma \Delta t}{(0.8)\left(\frac{4186 \text{ J}}{1 \text{ kcal}}\right)}$$

$$= \frac{((303 \text{ K})^4 - (293 \text{ K})^4)(1)\left(5.6704 \times 10^{-8} \frac{\text{W}}{\text{m}^2 \cdot \text{K}^4}\right)(2.1 \text{ m}^2)\left(24 \text{ h} \times \frac{3600 \text{ s}}{1 \text{ h}}\right)}{(0.8)\left(\frac{4186 \text{ J}}{1 \text{ kcal}}\right)}$$

$$= \boxed{3.2 \times 10^3 \text{ kcal}}$$

A value of 3.2×10^3 kcal seems $\boxed{\text{reasonable}}$ for a man who is 1.88 m (6 ft 2 in.) tall with a mass of 80 kg (176 lb).

REFLECT

The recommended caloric intake for an adult male is around 2550 kcal. It would make sense that additional, more complicated processes are occurring in the body than our simple model takes into account.

Chapter 15
Thermodynamics II

Conceptual Questions

15.1   Part a) Isothermal processes are those in which the temperature remains constant throughout. An example of an isothermal process is when an acid is slowly poured into a base and allowed to equilibrate with a water bath. The work done in isothermal processes (for an ideal gas) is given by $nRT\ln\left(\dfrac{V_f}{V_i}\right)$.

Part b) Adiabatic processes are those in which there is no heat lost or gained by the system. An example of an adiabatic process is a bicycle pump that is used to increase the pressure and temperature of the air that is inside as it goes into the tire. The work done in an adiabatic process is simply the change in internal energy (because $Q = 0$).

Part c) Isobaric processes are those in which the pressure remains constant. An example of a system undergoing an isobaric process is a gas in a cylinder that is sealed with a movable piston. As the gas is heated, the volume will expand to balance out the increase in temperature but the pressure remains constant.

Part d) Isochoric processes are those in which the volume does not change. An example of a system undergoing an isochoric process is a gas inside a cylinder that cannot expand or contract. If you change the temperature of the gas, it will respond by changing the pressure. No work is done in isochoric processes.

   Get Help: Picture It – P-V Diagram
   Interactive Exercise – Thermodynamics I
   Interactive Exercise – Thermodynamics II

15.5   Kinetic friction saps bulk kinetic energy and converts it into heat. Unless the engine was attempting to generate heat, the heat from friction is energy that won't be going toward whatever the engine was supposed to accomplish.

   Get Help: Picture It – Carnot Engine
   Interactive Exercise – Nuclear Power Plant

15.9   It increases by the amount of the heat divided by the temperature for the cool object minus the amount of heat divided by the temperature for the hot object: $\Delta S = \dfrac{Q}{T_c} - \dfrac{Q}{T_h}$.

When there was a temperature difference, that difference could be harnessed to do work. Once the two objects have equilibrated, that method of extracting work is no longer possible because the temperature difference no longer exists.

15.13 The irreversible processes are a bit like one-way streets—you can go around the block but not take a U-turn. More concretely, general heat loss is undone not by letting the heat back into the hot object but by finding something even hotter or by doing work on the object to heat it; the halting of a ball that rolled to a stop is not undone by letting the grass kick it back into motion but by going and fetching the ball. Note that the reversible adiabatic cooling does not let heat escape to the environment but rather stores the thermal energy as mechanical energy in another part of the system.

        Get Help: Picture It – Carnot Engine
                  Interactive Exercise – Nuclear Power Plant

15.17 As the piston moves, in principle the gas does work $p\Delta V$ with each change ΔV in volume, and heat ΔQ flows into the system from the reservoir to keep the internal energy of the ideal gas, and therefore its temperature, constant. In practice small temperature differences must exist to produce the flow of heat from the reservoir, so that the system deviates from the thermodynamic equilibrium. The more slowly the expansion occurs, the closer the behavior can be to ideal. Because the ideal behavior cannot be achieved, the entropy of the universe increases.

        Get Help: Picture It – Carnot Engine
                  Interactive Exercise – Nuclear Power Plant

Multiple-Choice Questions

15.21 B (decreases). $Q = 0$ because it is thermally isolated. Work is being done by the gas on the piston, which means the internal energy and, thus, the temperature of the gas decreases.

15.25 A (pressure). An isobaric process is one in which the pressure remains constant throughout.

        Get Help: Picture It – P-V Diagram
                  Interactive Exercise – Thermodynamics I
                  Interactive Exercise – Thermodynamics II

15.29 A (two adiabatic processes and two isothermal processes). The processes are isothermal expansion, adiabatic expansion, isothermal compression, and adiabatic compression.

        Get Help: Picture It – Carnot Engine
                  Interactive Exercise – Nuclear Power Plant
                  P'Cast 15.7 – Carnot Efficiency and Actual Efficiency

Estimation/Numerical Analysis

15.33 If we treat the gas as ideal, then $U = \frac{3}{2}nRT$. One liter of oxygen is about 0.045 mol, since 22.4 L = 1 mol of gas. At $T = 20°C$ (or 293 K),

$U_i = \frac{3}{2}(0.045 \text{ mol})\left(8.31\frac{\text{J}}{\text{mol} \cdot \text{K}}\right)(293 \text{ K}) \approx 164$ J and at $T = 100°C$ (or 373 K),

$U_f = \frac{3}{2}(0.045 \text{ mol})\left(8.31\frac{\text{J}}{\text{mol} \cdot \text{K}}\right)(373 \text{ K}) \approx 209$ J. Therefore, the change in internal energy is about 45 K.

15.37 Assume a 20-gallon tank and engine efficiency of 30%. The energy available to move the car is:

$$Q = (0.30)(20 \text{ gal})\left(\frac{125{,}000 \text{ BTU}}{\text{gal}}\right)\left(\frac{1054 \text{ J}}{\text{BTU}}\right) = 7.9 \times 10^8 \text{ J}$$

Rolling kinetic friction between the tires and the road dissipates energy from the car in the form of heat. The kinetic friction is related to the weight of the car by $f_k = \mu_k N = \mu_k mg$. The work done by this force on a 1200-kg car is:

$$W = Fd = f_k d = \mu_k mgd = Q$$

Solving for the distance:

$$d = \frac{Q}{\mu_k mg} = \frac{7.9 \times 10^8 \text{ J}}{(0.02)(1200 \text{ kg})\left(9.80\frac{\text{m}}{\text{s}^2}\right)} = 3{,}400{,}000 \text{ m} = 3400 \text{ km}$$

This answer is unrealistic, which suggests that there are other mechanical losses within the car itself, such as air drag, which would reduce this distance to a more realistic range.

Get Help: Picture It – Carnot Engine
Interactive Exercise – Nuclear Power Plant
P'Cast 15.6 – Efficiency of an Engine

Problems

15.45

SET UP

A gas is heated such that it follows a vertical line path on a pV diagram. The initial pressure and volume are 1.0×10^5 Pa and 3.0 m³, respectively. The final pressure and volume are 2.0×10^5 Pa and 3.0 m³, respectively. The work done by the gas on its surroundings is equal to zero because the gas neither expands nor contracts (that is, its volume remains constant).

SOLVE

$$W = 0 \text{ because } \Delta V = 0$$

REFLECT

Another way of thinking about this is that the area under a vertical line is zero. Since the volume remains constant throughout the process, this is an isochoric process.

Get Help: Picture It – P-V Diagram
Interactive Exercise – Thermodynamics I
Interactive Exercise – Thermodynamics II

166   Chapter 15 Thermodynamics II

15.47

SET UP

An expanding ideal gas does 8.8 kJ of work. The process occurs isothermally, which means the change in the internal energy of the gas is zero. From the first law of thermodynamics, this means the heat absorbed by the gas is equal to the work done by the gas.

SOLVE

$$\Delta U = 0 = Q - W$$

$$Q = W = \boxed{+8.8 \text{ kJ}}$$

REFLECT

The work done by a gas on its surroundings is considered to be positive, as is the heat absorbed by a gas.

Get Help: Picture It – P-V Diagram
Interactive Exercise – Thermodynamics I
Interactive Exercise – Thermodynamics II

15.51

SET UP

An ideal gas is contained in a vessel with fixed walls. While its pressure is decreased, 800 kJ of heat leaves the gas. Since the volume is fixed, the work done by the gas is equal to zero. The change in the internal energy of the gas is equal to Q in this case from the first law of thermodynamics. Heat leaving the system corresponds to a negative value, so $Q = -800$ kJ.

SOLVE

$$\Delta U = Q - W = (-800 \text{ kJ}) - 0 = \boxed{-800 \text{ kJ}}$$

REFLECT

This is an isochoric process because the volume is constant. Since the pressure of the gas decreases while the volume remains constant, the temperature of the gas must decrease. Therefore, we would expect heat to leave the system.

Get Help: Picture It – P-V Diagram
Interactive Exercise – Thermodynamics I
Interactive Exercise – Thermodynamics II

15.57

SET UP

The temperature of 4.00 g of helium gas is increased by 1.00 K at constant volume. We want to know the mass of oxygen gas that, using the same amount of heat, will experience a temperature increase of 1.00 K as well. The heat required to change the temperature of a gas at constant volume is given by $Q = nC_V\Delta T$. Using this expression, we can set up a ratio between the heats required by each gas, which are equal in this case, in terms of the numbers of moles and molar heat capacities at constant volume. The number of moles is equal to the mass divided by the molar mass of the gas. The data we'll need are the molar masses and the molar heat capacities at constant volume for both helium and oxygen: $M_{He} = 4.00 \frac{g}{mol}$, $M_{O_2} = 32.0 \frac{g}{mol}$, $C_{V,He} = 12.5 \frac{J}{mol \cdot K}$, and $C_{V,O_2} = 21.1 \frac{J}{mol \cdot K}$.

SOLVE

$$Q = nC_V \Delta T$$

$$\frac{Q_{O_2}}{Q_{He}} = \frac{n_{O_2} C_{V,O_2} \Delta T}{n_{He} C_{V,He} \Delta T}$$

$$1 = \frac{\left(\frac{m_{O_2}}{M_{O_2}}\right) C_{V,O_2}}{\left(\frac{m_{He}}{M_{He}}\right) C_{V,He}} = \left(\frac{m_{O_2}}{m_{He}}\right)\left(\frac{M_{He}}{M_{O_2}}\right)\left(\frac{C_{V,O_2}}{C_{V,He}}\right)$$

$$m_{O_2} = m_{He}\left(\frac{M_{O_2}}{M_{He}}\right)\left(\frac{C_{V,He}}{C_{V,O_2}}\right) = (4.00 \text{ g}) \frac{\left(32.0 \frac{g}{mol}\right)\left(12.5 \frac{J}{mol \cdot K}\right)}{\left(4.00 \frac{g}{mol}\right)\left(21.1 \frac{J}{mol \cdot K}\right)} = \boxed{19.0 \text{ g}}$$

REFLECT

This is about 0.6 mol of oxygen. For constant heat and temperature change, the product of the number of moles and the molar heat capacity is constant. Since the molar heat capacity of oxygen is larger than that of helium, the number of moles of oxygen must be less than the number of moles of helium under these circumstances.

15.61

SET UP

The temperature of air increases from $T_i = 300$ K to $T_f = 1130$ K. An ideal gas that undergoes an adiabatic expansion or compression obeys $TV^{\gamma-1}$ = constant. We can use this expression and $\gamma_{air} = 1.4$ to solve for the ratio of initial volume to final volume.

SOLVE

$$TV^{\gamma-1} = \text{constant}$$

$$T_i V_i^{\gamma-1} = T_f V_f^{\gamma-1}$$

$$\frac{T_f}{T_i} = \frac{V_i^{\gamma-1}}{V_f^{\gamma-1}} = \left(\frac{V_i}{V_f}\right)^{\gamma-1}$$

$$\frac{V_i}{V_f} = \left(\frac{T_f}{T_i}\right)^{\frac{1}{\gamma-1}} = \left(\frac{1130 \text{ K}}{300 \text{ K}}\right)^{\frac{1}{1.4-1}} = \boxed{27.5}$$

REFLECT

The final temperature is larger, so the adiabatic process will be compression. Therefore, $V_f < V_i$.

15.63

SET UP

An engine takes in 10 kJ and exhausts 6 kJ. The efficiency of the engine is equal to the difference in the energy it takes in and the energy it exhausts divided by the amount of energy it takes in: $e = \dfrac{W}{Q_H} = \dfrac{Q_H - |Q_C|}{Q_H}$.

SOLVE

$$e = \dfrac{W}{Q_H} = \dfrac{Q_H - |Q_C|}{Q_H} = \dfrac{(10 \text{ kJ}) - (6 \text{ kJ})}{10 \text{ kJ}} = 0.4 = \boxed{40\%}$$

REFLECT

Efficiency is most commonly given as a percentage.

          Get Help: Picture It – Carnot Engine
                     Interactive Exercise – Nuclear Power Plant
                     P'Cast 15.6 – Efficiency of an Engine

15.67

SET UP

A furnace supplies 28.0 kW of thermal power at $T_H = 573$ K and exhausts energy at $T_C = 293$ K. The theoretical maximum efficiency of a heat engine is given by $e_{\text{Carnot}} = 1 - \dfrac{T_C}{T_H}$. The actual efficiency is equal to the work output divided by the heat input, $e = \dfrac{W}{Q_H}$. We can divide the numerator and denominator by time, to convert the work and heat to work per second and thermal power, respectively. Assuming the actual efficiency is equal to the theoretical maximum efficiency, we can solve for the work per second $\dfrac{W}{\Delta t}$ given $\dfrac{Q_H}{\Delta t} = 28$ kW.

SOLVE

Theoretical maximum efficiency:

$$e_{\text{Carnot}} = 1 - \dfrac{T_C}{T_H} = 1 - \dfrac{293 \text{ K}}{573 \text{ K}} = 0.4887$$

Work per second:

$$e = \dfrac{W}{Q_H} = \dfrac{\dfrac{W}{\Delta t}}{\dfrac{Q_H}{\Delta t}}$$

We assume that $e = e_{\text{Carnot}}$.

$$\dfrac{W}{\Delta t} = e\left(\dfrac{Q_H}{\Delta t}\right) = (0.4887)(28.0 \text{ kW}) = \boxed{13.7 \text{ kW}}$$

REFLECT

This is the largest amount of work that can be expected since we've assumed the actual efficiency is equal to the theoretical maximum efficiency.

Get Help: Picture It – Carnot Engine
Interactive Exercise – Nuclear Power Plant
P'Cast 15.6 – Efficiency of an Engine

15.71

SET UP

A refrigerator requires $W = 35.0$ J to remove $Q_C = 190$ J from its interior. The temperature of the surroundings is $T_H = 295$ K. The coefficient of performance of the refrigerator is given by $\text{CP} = \dfrac{|Q_C|}{|W|}$. We can then use the other expressions for the coefficient of performance, $\text{CP} = \dfrac{|Q_C|}{|Q_H| - |Q_C|}$ and $\text{CP}_{\text{Carnot}} = \dfrac{T_C}{T_H - T_C}$, to calculate heat ejected to the surroundings Q_H and the internal temperature of the refrigerator, respectively.

SOLVE

Part a)

$$\text{CP} = \frac{|Q_C|}{|W|} = \frac{190 \text{ J}}{35.0 \text{ J}} = \boxed{5.4}$$

Part b)

$$\text{CP} = \frac{|Q_C|}{|Q_H| - |Q_C|}$$

$$(\text{CP})(|Q_H| - |Q_C|) = |Q_C|$$

$$|Q_H| = \frac{|Q_C| + (\text{CP})|Q_C|}{\text{CP}} = |Q_C|\left(\frac{1}{\text{CP}} + 1\right) = (190 \text{ J})\left(\frac{1}{5.4} + 1\right) = \boxed{2.3 \times 10^2 \text{ J}}$$

Part c)

$$\text{CP}_{\text{Carnot}} = \frac{T_C}{T_H - T_C}$$

$$(\text{CP}_{\text{Carnot}})(T_H - T_C) = T_C$$

$$T_C = \frac{(\text{CP}_{\text{Carnot}})T_H}{1 + \text{CP}_{\text{Carnot}}} = \frac{(5.4)(295 \text{ K})}{1 + 5.4} = 249 \text{ K} = \boxed{-24°\text{C}}$$

REFLECT

The fact that the refrigerator cycle is reversible means it follows the Carnot cycle, which allows us to use the expression for the coefficient of performance based on the temperatures.

Get Help: Picture It – Carnot Engine
Interactive Exercise – Nuclear Power Plant

15.73

SET UP

A sample of ice absorbs 6.68×10^4 J of heat at 273 K as it melts. The minimum entropy change of the system is given by $\Delta S = \dfrac{Q}{T}$.

SOLVE

$$\Delta S = \frac{Q}{T} = \frac{6.68 \times 10^4 \text{ J}}{273 \text{ K}} = \boxed{245 \frac{\text{J}}{\text{K}}}$$

REFLECT

The entropy change is equal to $\dfrac{Q}{T}$ only in the case of a reversible process. We do not need to use the mass of the ice since we are given the amount of heat absorbed.

15.77

SET UP

A car (m_{car} = 1800 kg) is traveling at a speed of v_{ix} = 80.0 km/h when it crashes into a wall. The temperature of the air is 300.2 K. The entropy change of the universe due to an irreversible process is equal to the total energy transferred divided by the temperature, $\Delta S = \dfrac{Q}{T}$.

SOLVE

$$E_{car} = K_i = \frac{1}{2}m_{car}v_{ix}^2$$

Assuming all of the car's energy is transferred as heat:

$$E_{car} = Q = \frac{1}{2}m_{car}v_i^2$$

$$\Delta S = \frac{Q}{T} = \frac{\left(\frac{1}{2}m_{car}v_{iy}^2\right)}{T} = \frac{\left(\frac{1}{2}\right)(1800 \text{ kg})\left(80\dfrac{\text{km}}{\text{h}} \times \dfrac{1000 \text{ m}}{1 \text{ km}} \times \dfrac{1 \text{ h}}{3600 \text{ s}}\right)^2}{300.2 \text{ K}} = \boxed{1.5 \times 10^3 \frac{\text{J}}{\text{K}} = 1.5 \frac{\text{kJ}}{\text{K}}}$$

REFLECT

We're assuming that all of the initial kinetic energy of the car is absorbed by the universe as heat.

15.81

SET UP

In an incompletely insulated bottle, water (m_{water} = 0.35 kg) and ice (m_{ice} = 0.15 kg) are initially at equilibrium at 273.15 K. Over time, the contents of the bottle come to thermal equilibrium with the outside air temperature, T_{air} = 298.15 K. Heat is transferred from the air to the ice and water—first, the ice melts and then all of the water heats up to 298.15 K. The heat necessary to melt the ice is related to its latent heat of fusion $\left(L_F = 334 \times 10^3 \dfrac{\text{J}}{\text{kg}}\right)$;

the heat necessary to increase the temperature of the water is related to its specific heat $\left(C_{\text{water}} = 4186 \dfrac{\text{J}}{\text{kg} \cdot \text{K}}\right)$. The entropy changes of the melting ice and air are equal to $\Delta S_{\text{melting}} = \dfrac{Q_{\text{melting}}}{T_{\text{ice}}}$ and $\Delta S_{\text{air}} = \dfrac{Q_{\text{air}}}{T_{\text{air}}}$, respectively. The total change in entropy of the universe is the sum of these three entropy changes.

SOLVE
Heat transferred:
$$Q = mc\Delta T$$

$$Q_{\text{water}} = m_{\text{water}} C_{\text{water}} (T_{\text{eq}} - T_i) = (0.35 \text{ kg})\left(4186 \dfrac{\text{J}}{\text{kg} \cdot \text{K}}\right)((298.15 \text{ K}) - (273.15 \text{ K}))$$
$$= 36{,}627.5 \text{ J}$$

$$Q_{\text{ice}} = Q_{\text{melting}} + Q_{\text{warming}} = m_{\text{ice}} L_F + m_{\text{ice}} C_{\text{water}}(T_{\text{eq}} - T_i) = m_{\text{ice}}(L_F + C_{\text{water}}(T_{\text{eq}} - T_i))$$
$$= (0.15 \text{ kg})\left(\left(334 \times 10^3 \dfrac{\text{J}}{\text{kg}}\right) + \left(4186 \dfrac{\text{J}}{\text{kg} \cdot \text{K}}\right)((298.15 \text{ K}) - (273.15 \text{ K}))\right) = 65{,}797.5 \text{ J}$$

$$Q_{\text{air}} = -(Q_{\text{water}} + Q_{\text{ice}}) = -((36{,}627.5 \text{ J}) + (65{,}797.5 \text{ J})) = -102{,}425 \text{ J}$$

Entropy change of the ice melting:
$$\Delta S = \dfrac{Q}{T}$$

$$\Delta S_{\text{melting}} = \dfrac{Q_{\text{melting}}}{T_{\text{ice}}} = \dfrac{m_{\text{ice}} L_F}{T_{\text{ice}}} = \dfrac{(0.15 \text{ Kg})\left(334 \times 10^3 \dfrac{\text{J}}{\text{Kg}}\right)}{273.15 \text{ K}} = \dfrac{50{,}100 \text{ J}}{273.15 \text{ K}} = 183.4 \dfrac{\text{J}}{\text{K}}$$

Entropy change of the water warming:
$$\Delta S_{\text{warming}} = m C_{\text{water}} \ln\left(\dfrac{T_{\text{eq}}}{T_i}\right)$$
$$= ((0.15 \text{ kg}) + (0.35 \text{ kg}))\left(4186 \dfrac{\text{J}}{\text{kg} \cdot \text{K}}\right) \ln\left(\dfrac{298.15 \text{ K}}{273.15 \text{ K}}\right) = 183.3 \dfrac{\text{J}}{\text{K}}$$

Entropy change of the air:
$$\Delta S_{\text{air}} = \dfrac{Q_{\text{air}}}{T_{\text{air}}} = \dfrac{-102{,}529 \text{ J}}{298.2 \text{ K}} = -343.8 \dfrac{\text{J}}{\text{K}}$$

Entropy change of the universe:
$$\Delta S = \Delta S_{\text{melting}} + \Delta S_{\text{warming}} + \Delta S_{\text{air}} = \left(183.4 \dfrac{\text{J}}{\text{K}}\right) + \left(183.3 \dfrac{\text{J}}{\text{K}}\right) + \left(-343.8 \dfrac{\text{J}}{\text{K}}\right) = \boxed{23 \dfrac{\text{J}}{\text{K}}}$$

REFLECT
We can only use $\Delta S = \dfrac{Q}{T}$ if the temperature is constant.

General Problems

15.83

SET UP

A Carnot engine extracts heat Q_H from seawater (T_H = 291 K) to power a ship. The exhausted heat Q_C is used to sublimate a reserve of dry ice of mass m at a temperature T_C = 195 K. The ship's engines run at 8000 horsepower, where 1 hp = 746 W, for a full day; multiplying the engine's power by the time will give the work the engine is required to output to run the ship. We can use this value, the efficiency of a Carnot engine, and the fact that $|W| = |Q_H| - |Q_C|$ to find the amount of heat Q_C that will be exhausted. All of the exhausted heat will go to sublimating the dry ice, which means $Q_C = mL_F$; the latent heat of sublimation of carbon dioxide is $L_S = 573,700 \frac{J}{kg}$.

SOLVE

Energy required for one day:

$$W = P\Delta t = \left(8000 \text{ hp} \times \frac{746 \text{ W}}{1 \text{ hp}}\right)\left(1 \text{ d} \times \frac{24 \text{ h}}{1 \text{ d}} \times \frac{3600 \text{ s}}{1 \text{ h}}\right) = 5.156 \times 10^{11} \text{ J}$$

Efficiency of the engine:

$$e_{\text{Carnot}} = 1 - \frac{T_C}{T_H} = 1 - \frac{195 \text{ K}}{291 \text{ K}} = 0.3299$$

Total heat exhausted:

$$e_{\text{Carnot}} = \frac{W}{Q_H} = \frac{W}{W + |Q_C|} = \frac{W}{W + |mL_S|}$$

$$m = \frac{|W|(1 - e_{\text{Carnot}})}{e_{\text{Carnot}}|L_S|} = \frac{(5.156 \times 10^{11} \text{ J})(1 - 0.3299)}{(0.3299)\left(573,700 \frac{J}{kg}\right)} = \boxed{1.83 \times 10^6 \text{ kg}}$$

REFLECT

An engine uses heat to do work; a refrigerator does work to remove heat.

          Get Help: Picture It – Carnot Engine
                     Interactive Exercise – Nuclear Power Plant
                     P'Cast 15.7 – Carnot Efficiency and Actual Efficiency

15.87

SET UP

When a 60.0-kg woman walks up five flights of stairs (Δh = 20.0 m), she normally releases 100 kJ of heat; if she has a high fever, she gives off 10% more heat. The work done by the person to climb the stairs is equal to the change in her gravitational potential energy. The efficiency of the woman's body is equal to the work done divided by the total energy she outputs, which is equal to the work done plus the heat emitted.

SOLVE

Work done:

$$W = \Delta U = mg\Delta h = (60.0 \text{ kg})\left(9.80\frac{\text{m}}{\text{s}^2}\right)(20.0 \text{ m}) = 11{,}760 \text{ J} = 11.76 \text{ kJ}$$

Normal temperature:

$$Q_C = 100 \text{ kJ}$$

$$|Q_H| = |W| + |Q_C| = 11.76 \text{ kJ} + 100 \text{ kJ} = 111.76 \text{ kJ}$$

$$e = \frac{W}{Q_H} = \frac{11.76 \text{ kJ}}{111.76 \text{ kJ}} = \boxed{0.105 = 10.5\%}$$

Elevated temperature:

$$Q_C = 1.1(100 \text{ kJ}) = 110 \text{ kJ}$$

$$|Q_H| = |W| + |Q_C| = 11.76 \text{ kJ} + 110 \text{ kJ} = 121.76 \text{ kJ}$$

$$e = \frac{W}{Q_H} = \frac{11.76 \text{ kJ}}{121.76 \text{ kJ}} = \boxed{0.0966 = 9.66\%}$$

The patient's $\boxed{\text{efficiency drops when she has a fever}}$.

REFLECT

Since the patient's efficiency is lower when she is sick, it will take her more energy to accomplish the same tasks she did when healthy.

Get Help: Picture It – Carnot Engine
Interactive Exercise – Nuclear Power Plant
P'Cast 15.6 – Efficiency of an Engine

15.91

SET UP

The temperature increases by 3.0°C for every 100 m we drill into Earth's crust. Water pumped into an oil well that is 1830 m deep is used as a heat engine. The temperature of the hot reservoir (that is, the bottom of the well) is equal to the depth multiplied by the relationship between the temperature increase and depth. The temperature of the cold reservoir (that is, the surface) is $T_C = 293$ K. The maximum efficiency occurs when the heat engine acts reversibly, so $e_{\text{Carnot}} = 1 - \frac{T_C}{T_H}$. We want to use a combination of these wells to create a reversible power plant that provides 2.5×10^6 W of power. We can use a different expression for efficiency, $Q_H = \frac{W}{e}$, in order to calculate the total heat the power plant absorbs from Earth in one day.

SOLVE

Part a)

Maximum temperature:

$$T_H = 20°C + (1830 \text{ m})\left(\frac{3.0°C}{100 \text{ m}}\right) = 74.9°C = 348 \text{ K}$$

Maximum efficiency:

$$e_{\text{Carnot}} = 1 - \frac{T_C}{T_H} = 1 - \frac{293 \text{ K}}{348 \text{ K}} = 0.158 = 16\%$$

Part b)

$$e = \frac{W}{Q_H}$$

$$Q_H = \frac{W}{e} = \frac{(2.5 \times 10^6 \text{ W})\left(1 \text{ d} \times \frac{24 \text{ h}}{1 \text{ d}} \times \frac{3600 \text{ s}}{1 \text{ h}}\right)}{0.158} = \boxed{1.4 \times 10^{12} \text{ J} = 1.4 \text{ TJ}}$$

REFLECT

A terajoule (TJ) is equal to 1×10^{12} J. Because we used the maximum efficiency, 1.4 TJ is the minimum possible energy that is absorbed from the interior of Earth in one day. If the actual efficiency is smaller than 16%, the power plant must absorb more than 1.4 TJ of heat in a day in order to deliver a power of 2.5 MW.

Get Help: Picture It – Carnot Engine
Interactive Exercise – Nuclear Power Plant
P'Cast 15.6 – Efficiency of an Engine

15.95

SET UP

A heat engine absorbs $Q_H = 1250$ J from a high-temperature reservoir ($T_H = 490$ K) and does 475 J of work. The temperature of the cold reservoir is $T_C = 273$ K. The efficiency of the engine is equal to the work done divided by the heat absorbed. The change in the entropy of the universe is equal to the sum of the entropy changes of the hot and cold reservoirs. The heat exhausted to the cold reservoir Q_C was not given but is equal to $Q_H - W$ from the first law of thermodynamics. Finally, the amount of energy unavailable for doing work after one full cycle is equal to $T_C \Delta S$.

SOLVE

Part a)

$$e = \frac{W}{Q_H} = \frac{475 \text{ J}}{1250 \text{ J}} = \boxed{0.38 = 38\%}$$

Part b)

$$|W| = |Q_H| - |Q_C|$$

$$|Q_C| = |Q_H| - |W|$$

$$\Delta S = \Delta S_H + \Delta S_C = \frac{Q_H}{T_H} + \frac{Q_C}{T_C} = \frac{Q_H}{T_H} + \frac{Q_H - W}{T_C}$$

$$= \frac{-1250 \text{ J}}{490 \text{ K}} + \frac{(1250 \text{ J}) - (475 \text{ J})}{273 \text{ K}} = \boxed{0.288 \frac{\text{J}}{\text{K}}}$$

Part c)

$$W_{unavailable} = T_C \Delta S = (273 \text{ K})\left(6.7482 \frac{\text{J}}{\text{K}}\right) = \boxed{1.84 \times 10^3 \text{ J}}$$

REFLECT

The amount of energy unavailable to do work is also equal to the difference between the maximum work possible in a reversible process and the actual work done by a system.

15.99

SET UP

A certain person typically eats about 2250 kcal per day, 80.0% of which is transferred to the surroundings as heat. The surroundings are at 295.15 K. The entropy change of the surroundings is equal to the heat absorbed divided by the temperature.

SOLVE

$$\Delta S = \frac{Q}{T} = \frac{(0.800)\left(2250 \times 10^3 \text{ cal} \times \frac{4.184 \text{ J}}{1 \text{ cal}}\right)}{295.15 \text{ K}} = \boxed{2.55 \times 10^4 \frac{\text{J}}{\text{K}} = 25.5 \frac{\text{kJ}}{\text{K}}}$$

The entropy of the apartment $\boxed{\text{increases}}$.

REFLECT

The surroundings absorb heat, so the entropy of the surroundings should increase.